Clinical Applications of Continuous INFUSION CHEMOTHERAPY and Concomitant RADIATION THERAPY

Clinical Applications of Continuous INFUSION CHEMOTHERAPY and Concomitant RADIATION THERAPY

Edited by
C. Julian Rosenthal, M.D.
*Associate Professor of Medicine
Director, Division of Medical Oncology
State University of New York
Health Science Center at Brooklyn
Brooklyn, New York*

and
Marvin Rotman, M.D.
*Professor and Chairman
Department of Radiation Oncology
State University of New York
Health Science Center of Brooklyn
Brooklyn, New York*

SPRINGER SCIENCE+BUSINESS MEDIA, LLC

Library of Congress Cataloging in Publication Data

Conference on Continuous Infusion Chemotherapy and its Interactions with Radiation in the Treatment of Malignant Tumors (1st: 1985: New York, N.Y.)
Clinical applications of continuous infusion chemotherapy and concomitant radiation therapy.

"Proceedings of the First Conference on Continuous Infusion Chemotherapy and its Interactions with Radiation in the Treatment of Malignant Tumors, held March 22, 1985, in New York, New York"—T.p. verso.
Includes bibliographies and index.
1. Cancer—Adjuvant treatment—Congresses. 2. Antineoplastic agents—Congresses. 3. Cancer—Radiotherapy—Congresses. 4. Drugs—Administration—Congresses. I. Rosenthal, C. J. (C. Julian) II. Rotman, Marvin, 1933-   . III. Title. IV. Title: Continuous infusion chemotherapy and concomitant radiation therapy. [DNLM: 1. Antineoplastic Agents—administration & dosage—congresses. 2. Infusions, Parenteral—methods—congresses. 3. Neoplasms—drug therapy—congresses. 4. Neoplasms—radiotherapy—congresses. W3 C572N 1st 1985c / QZ 267 C7472 1985c]
RC271.A35C66   1985   616.99'406   86-4890
ISBN 978-1-4612-9291-3   ISBN 978-1-4613-2197-2 (eBook)
DOI 10.1007/ 978-1-4613-2197-2

Proceedings of the First Conference on Continuous Infusion Chemotherapy
and its Interactions with Radiation in the Treatment of Malignant Tumors,
held March 22, 1985, in New York, New York

© 1986 Springer Science+Business Media New York
Originally published by Plenum Press, New York in 1986
Softcover reprint of the hardcover 1st edition 1986

All rights reserved

No part of this book may be reproduced, stored in a retrieval system, or transmitted
in any form or by any means, electronic, mechanical, photocopying, microfilming,
recording, or otherwise, without written permission from the Publisher

PREFACE

The first conference on concomitant infusion chemotherapy and radiation therapy was organized with the intention of bringing together some of the investigators who have tested, during the last few years, the hypothesis that continuous infusion chemotherapy could modulate the cytotoxic effect of radiation therapy to the point of having a strongly additive, if not synergistic activity on certain malignant tumors.

This volume represents the detailed proceedings of this conference presented in a way that offers the reader a review of the on-going research in the field. We have stressed a number of subjects from basic biologic research and influence of cell kinetics to the practical methods of drug delivery systems and early clinical experiences.

The rationale for this new type of combined modality therapy has been presented by some of its pioneers. Early clinical investigations as well as the preliminary data of many that have not yet completely matured have also been included. The reader should look at these data with some reservations. Ultimately, these results must be confirmed by larger prospective randomized studies with proper controls before becoming accepted as the treatment of choice in locally advanced tumors.

It is our hope that this volume will succeed in bringing to the interested medical community a new approach in clinical medicine which, in paraphrasing Konrad Lorenz can be considered as "the working hypothesis best suited to open the way to the next better one", aiming ultimately at the ablation of localized malignant tumors with a minimum of morbidity.

C. Julian Rosenthal
Marvin Rotman

CONTENTS

SECTION I

PROTRACTED DELIVERY OF ANTINEOPLASTIC CHEMOTHERAPY AGENTS

A. Principles and Therapeutic Applications

Theoretical, Clinical and Pharmacokinetic Aspects
    of Cancer Chemotherapy Administered by
    Continuous Infusion   3
  B.I. Sikic

Pharmacology and Therapeutic Efficacy of Bleomycin
    Administered by Continuous Infusion 13
  I.H. Krakoff

Continuous-Infusion Adriamycin 19
  R.S. Benjamin, S.P. Chowla, G.N. Hortobagyi, M.S. Ewer,
  B. Mackay, S.S. Legha, H. Carrasco, and S. Wallace

5-Fluorouracil Plus Thymidine or Leucovorin by
    Continuous i.v. Infusion in the Treatment of
    Advanced Colorectal Carcinoma 27
  F. Trave and Y.M. Rustum

Epipodophyllotoxin and Cisplatin on Continuous
    Infusion Schedules 43
  J.J. Lokich

Biodegradable Starch Microspheres (SPHEREX): A Clinical
    Useful Medical Device for Combined Intra-arterial
    Chemotherapeutic Treatment of Primary and
    Metastatic Cancers of the Liver 51
  G. Parker and W. Regelson

Selective Therapy of Hepatic Cancers Using Microspheres . . . 67
  J.W. Gyves

B. Clinical Studies

Preliminary Results of a Randomized Study of Intrahepatic
    Infusion Versus Systemic Infusion of FUDR
    for Metastatic Colorectal Carcinoma 73
  N. Kemeny and J. Daly

COPBLAM: Infusion Chemotherapy for Large Cell Lymphoma 79
   M. Coleman, B. Boyd, B. Bernhardt, G. Gerstein,
   and S. Kopel

Adriamycin Continuous i.v. Infusion for the Treatment
   of Childhood Hepatic Malignancies, Toxicity and
   Efficacy: A Pilot Study Childrens Cancer Study Group 87
   J.A. Ortega, W. Woods, J. Feusner, G. Reaman,
   B. Lange, and G.D. Hammond

An Uncontrolled Phase II Study of Constant Infusion
   Vincristine-Adriamycin 91
   L. Helson, M.A. Costello, E. Arenson,
   L. Steinherz, and S. Groshen

Low Dose Ara-C by Continuous Infusion in the Treatment
   of Acute Non-Lymphocytic Leukemia (ANLL) and
   Myelodysplastic Syndrome (MDS) 95
   F. Chaudri, S.L. Allen, P. Schulman, W. Kreis,
   D.R. Budman, L. Weiselberg, and V. Vinceguerra

The 5-Day Continuous Infusion of Cis-platinum: An
   Update on Toxicity Pattern 101
   P. Salem, M. Khalyl, K. Jabboury, and L. Hashimi

Long Term Appraisal of the Lemon-Foley Method of Chemotherapy
   of Solid Tumors of the Gastrointestinal Tract 107
   D. Steinberg

SECTION II

ANTINEOPLASTIC EFFECTS OF RADIATION THERAPY AND CONCOMITANT
CHEMOTHERAPY BY CONTINUOUS INFUSION

   A. Principles and Therapeutic Applications

Theoretical Basis and Clinical Applications of
   5-Fluorouracil as a Radiosensitizer 113
   J.E. Byfield

Treatment of Hepatic Metastases from Gastro Intestinal
   Primaries with Split Course Radiation Therapy
   and Concomitant Infusion 5-Flourouracil 127
   M. Rotman, I. Bhutiani, A. Kuruvilla, K. Choi,
   C.J. Rosenthal, A. Braverman and J. Marti

Combined Modality Therapy with 5-Fluorouracil, Mitomycin-C
   and Radiation Therapy for Squamous Cell Cancers 133
   B.J. Cummings, T.J. Keane, A.R. Harwood, and G.M. Thomas

Treatment of Bladder Carcinoma with Concomitant Infusion
   Chemotherapy and Irradiation 149
   M. Rotman, R. Macchia, M. Silverstein, K. Choi,
   C.J. Rosenthal, A. Braverman and M. Aziz

A Urologist's Viewpoint: Treatment of Invasive Bladder
   Cancer by the XRT/5FU Protocol 155
   R.J. Macchia and G. Langauni

Concomitant Radiation Therapy and Doxorubicin by
    Continuous Infusion in Advanced Malignancies - A
    Phase I-II Study - Evidence of Synergistic Effect
    in Soft Tissue Sarcomas and Hepatomas 159
  C.J. Rosenthal, M. Rotman, and I. Bhutiani

Cis Platin by Continuous Infusion with Concurrent
    Radiation Therapy in Malignant Tumors (A Phase I-II
    Study) 177
  C.J. Rosenthal, M. Rotman, K. Choi and J. Sand

B. Clinical Studies

Combination of Radiation with Concomitant Continuous
    Adriamycin Infusion in a Patient with Partially Excised
    Pleomorphic Soft Tissue Sarcoma of the Lower Extremity 183
  S. Turner, R. Shetty, H. Gandhi, A. Latyshevsky,
  J. Korzis and R. Yaes

Treatment of Recurrent Carcinoma of the Paranasal Sinuses
    Using Concomitant Infusion Cis-platinum And Radiation
    Therapy 189
  M. Rotman, K. Choi, S. Isaacson, C.J. Rosenthal,
  A. Braverman, and J. Marti

Hepatic Artery Infusion (HAI) for Hepatic Metastases in
    Combination with Hepatic Resection and Hepatic Radiation . . . 195
  R.W. Merrick, R.R. Dobelbower, Jr., J.F. Ringleiut,
  and R.T. Skeel

Phase II Study of Simultaneous Radiation Therapy, Continuous
    Infusion, 5-FU and Bolus Mitomycin-C 201
  O.A. Mendiondo, L.C. Maguire, W.D. Medina, and J.D. Cronin

Cancer of the Esophagus - Medical University of
    South Carolina Pilot Study 207
  J.M. Jenrette, III, R.D. Marks Jr., E.F. Parker, and
  R.H. Fitzgerald Jr.

Continuous Infusion VP-16, Bolus Cis-Platinum and
    Simultaneous Radiation Therapy as Salvage Therapy
    in Small Cell Bronchogenic Carcinoma 211
  K. Rowland Jr., P. Bonomi, S. Taylor, S. Maffey,
  S. Reddy, and M.S. Lee

Concomitant Radiation, Mitomycin-C and 5-Fluorouracil
    Infusion in Gastro Intestinal Cancer - A
    Preliminary Report 217
  A. Chan, A. Wong and K. Arthur

SECTION III

TECHNIQUES FOR THE ADMINISTRATION OF CHEMOTHERAPY AGENTS BY CONTINUOUS INFUSION

Procedures for the Use of Implantable and External Pumps
    for Continuous Infusion Chemotherapy 223
  B.B. Willis

Central Line Catheter Care: The Nurses' and Patients
    Perspective 229
  M.J. Tunny

Potential Complications of Right Atrial Catheterization 233
  R.M. Stillman

Long Term Complications of the Indwelling Central
    Line Catheters 239
  J. Marti

INDEX 243

INTRODUCTION

The history of cancer therapy has known, from its earliest descriptions in ancient Egyptian papyruses (Petrie, Smith, Ebbers) (1), centuries of frustration. It was not until the technical revolution of the late nineteenth century which led to the introduction of radiation therapy in the treatment of cancer was there a major therapeutic potential for erradication of malignant diseases. However, the massive single exposures of radiation and their accompanying high morbidity and local failure rate plunged this new medical discipline into a period of disillusionment and frustration. Only after Claude Regaud, in 1919, found that fractionated radiation directed to the testicle was capable of eliminating spermatagenesis without affecting the surrounding tissues was the treatment of cancer with radiation revived (2). This technique applied to the therapy of malignant tumors of the head and neck and cervix and improved disease control without the then high anticipated complication rate (3). The later technologic advances of kilovoltage and megavoltage succeeded to improve on the early results. Nevertheless, the incidence of local and regional disease failure remained high while other tumors such as the soft tissue sarcomas were seemingly unresponsive to radiation.

It has been close to half a century since initial radiobiological experiments pointed out the radiosensitizing effects of oxygen and the radio-resistance afforded by hypoxia. Since that time, efforts to overcome radio-resistance have been directed towards refining the treatment techniques, using the sophisticated plans of high energy machines. It is primarily with the improved understanding of tumor biology in the last two decades that attempts to overcome this hypoxic fraction of tumor cells had led to the use of heavy particle irradiation, hyperthermia, hyperfractionation and, in particular, to the investigation of radiosensitizers. It had been hoped that electron-affinic compounds would selectively radiosensitize the resistant poorly oxygenated tissues to the usual dose of radiation. The nitromidazol compounds and their analysis have as yet not shown the ability to radiosensitize tumor cells; innate toxicities have prevented their use in concentrations required for tumor response. The failure thus far of the electron-affinic compounds lead investigators to look at the concomitant use of radiation therapy and chemotherapy.

The evolution of chemotherapy contains events similar to those found in the history of radiation therapy. Initially, in the 1940's alkylating agents showed promise of disease erradication. The morbidity and lack of tumor sterilization prompted the continued production of new drugs for the much sought after cancer "cure". Soon methotrexate followed by corticosteroids and certain antibiotics (4) were found capable of inducing partial tumor response. When used in combination, these drugs produced prolonged complete remissions equivalent to cures, in some cases, of acute lymphocytic leukemias, advanced lymphomas and choriocarcinomas in females.

The discovery of new drugs over the next few decades (Bleomycin, Cisplatin, Etoposide) enlarged the percentages of response rates but did not significantly reduce recurrence rates.

Attempts to increase antineoplastic activity by concentrating the drugs to the tumor site have included their intracavitary or intra-arterial administration. In the past decade administration by continuous intravenous infusion was encouraged by experimental in-vitro data (5) which showed that the cytocidal effect of various antineoplastic chemotherapy agents related not only to concentration of drugs but to their duration of contact with the malignant cells.

This belief was reinforced by the clinical observations of Nigro (6), Byfield (7), Sischy (8), Cummings (9) and Rotman indicating that squamous cell carcinoma of the anus and esophagus, adenocarcinoma of the rectum and transitional cell carcinoma to the bladder would regress when exposed to infusion 5FU and radiotherapy. In addition, early phase II studies conducted by Rosenthal, Rotman, et. al. (10,11) suggested a synergistic antineoplastic effect of radiation therapy and continuous infusion adriamycin for selective soft tissue sarcomas and hepatomas as well as of infusion cisplatin and radiation therapy for some advanced squamous cell carcinomas (12). There are not, as yet, controlled prospective randomized studies to demonstrate the clinical benefits of the combined concomitant modalities versus radiation therapy alone. Despite this, the editors of this volume thought that there was enough merit in the pilot work on continuous infusion chemotherapy and irradiation to lead to the organization of the first conference on continuous infusion chemotherapy and concomitant radiation therapy.

The intention of the editors of this volume was not to exhaustively review all the pioneering clinical and basic scientific work which led to the current status of the concomitant use of radiation and infusion chemotherapy nor to include results of all the past or ongoing clinical trials in this field. Rather, this volume intends to bring to the attention of clinicians and clinical investigators, the most promising data of a new therapeutic approach in which we think significant progress in the therapy of malignant tumors will be registered and, hopefully, to raise the interest of many clinicians to this area of fruitful clinical research. In this respect, we would like to caution the readers against the casual application of the methods described in the various chapters. They should be applied only in centers where rigorous control of the techniques and appropriate supportive therapy are available and always within the framework of Internal Research Board approved investigative protocols. *The editors and the publisher cannot guarantee the accuracy of the drug dosage listed in various papers. No regimen listed in the various articles of this book should be administered without prior consultation with the author(s) of the article describing it.*

We would also like to acknowledge the invaluable secretarial help of Ms. Amrani and Ms. Dutka, the expert editorial help of Ms. Raynor-Enco and the support received from the American Cancer Society (grant #2484), National Cancer Institute (grant #2R25CA1798206) and the Bristol Company without which the conference and this volume on concomitant continuous infusion chemotherapy and radiation therapy would have not been possible.

September 18, 1985

C. Julian Rosenthal
Marvin Rotman

REFERENCES

1. Haagensen,C.D. An exhibit of important books, papers and memorabilia illustrating the condition of the knowledge of Cancer. Am. J. Cancer 18:42-126, 1933.
2. Bainbridge,W.S. The Cancer Problem. New York, MacMillan 1914, pp. 1-36, 106-129.
3. Case,J.T.: History of radiation therapy in Buschke, F. (ed.). Progr. in Radiation Therapy. Vol. 2. New York, Grune and Stratton 1958, pp. 13-41.
4. Zubrod,C.G. Historic Milestones in Curative Chemotherapy. Sem. Onc. 6:490-505, 1979.
5. Shimoyama,M. The cytocidal action of alkylating agents and anticancer antibiotics against in vitro cultured Yoshida ascites sarcoma cells. Jap. Soc. Cancer Ther. 10:63-72, 1975.
6. Nigro,N.D., Vaitkevicius,V.K. and Considine,B. Combined therapy for cancer of the anal canal: a preliminary report. Dis. Colon Rectum 17:354-356, 1974.
7. Byfield,J.E., Calabro Jones,P., Klisak,I. and Kulhanian,F. Pharmacologic requirements for obtaining sensitization of human tumor cells in vitro to combined 5-Fluorouracil or Ftoraful and x-rays. Int. J. Radiat. Oncol. Biol. Phys. 8:1923-1933, 1982.
8. Sischy,B., Qazi,R. and Hinson,E.J. A pilot study of concurrent radiation, Mitomycin C and 5-FU in marginally operable carcinoma of the rectum. Inter. J. Radiation Oncol. Biol. Phys. 10 (suppl. 2): 9 (abs.), 1984.
9. Cummings,B.J. and Byfield,J.E. Anal Cancer in "Innovations in Radiation Oncology Research" (H.R. Withers and L.J. Peters eds.) Springer-Verlag (in press).
10. Rosenthal,C., Bhutiani,I., Choi,K. and Rotman,M. Low dose Adriamycin by continuous intravenous infusion with and without concomitant radiation therapy. Proc. 13th Int. Cong. on Cancer; Abst. #3446:603, 1983.
11. Rosenthal,C.J., Bhutiani,I. and Rotman,M. Adriamycin by continuous infusion potentiates the effect of concurrent radiation therapy in soft tissue sarcomas and hepatomas. 13th Int. Cong. of Chemoth. 13: pp. 1-10, 1983.
12. Rosenthal,C.J., Rotman,M., Choi,K. and Sand,J. Cisplatin by continuous infusion with concomitant radiation therapy in malignant tumors (a phase I-II study). Proc. 14th Int. Cong. of Chemoth. Abst. #S-69-1, 14:135, 1985.

SECTION I: PROTRACTED ADMINISTRATION OF ANTINEOPLASTIC CHEMOTHERAPY AGENTS

    A. Principles and Therapeutic Applications

THEORETICAL, CLINICAL, AND PHARMACOKINETIC ASPECTS OF CANCER CHEMOTHERAPY ADMINISTERED BY CONTINUOUS INFUSION

Branimir Ivan Sikic

Department of Medicine
Oncology and Clinical Pharmacology
Stanford University School of Medicine

Stanford Medical Center, M-211
Stanford, CA 94305

INTRODUCTION

Most anti-cancer drugs produce toxic side-effects at the doses used for treatment. This low therapeutic index, or ratio between therapeutic and toxic doses, has prompted various approaches towards increasing the selectivity of cancer chemotherapy. Perhaps the most successful of these has been the use of combination chemotherapy, which increased the therapeutic ratio by combining drugs with non-overlapping side-effects and in some cases lowering individual doses. This approach over the last twenty years has resulted in the cure of several types of cancer (leukemias, Hodgkin's and non-Hodgkin's lymphomas, germ cell cancers). The use of drugs in combination is not curative for most human solid tumors, however, and in fact may increase toxicity without increasing therapeutic efficacy for tumors which are refractory to the drugs employed.

Other approaches to increasing selectivity have included regional perfusions (e.g., through the hepatic artery for hepatomas or isolated liver metastases), intra-cavitary administration (e.g., intra-peritoneal, large-volume drug administration for ovarian carcinomas), and the judicious use of antidotes (e.g., leucovorin with higher doses of methotrexate).

Alterations in the schedule of drug administration may also affect the efficacy and toxicity of anti-neoplastic agents. In particular, continuous intravenous infusion of some of these drugs may improve their therapeutic index compared to conventional intravenous bolus injection (1, 2). In this chapter, we will review some of the theoretical and empirical aspects of the administration of anti-cancer drugs by continuous intravenous infusion.

VARIABLES CONTRIBUTING TO SCHEDULE DEPENDENCE OF ANTI-CANCER DRUGS

Pharmacokinetics

Drugs which have a short plasma half-life, due to metabolic inactivation or excretion, tend to be more active by continuous infusion compared to bolus administration. For example, cytarabine is extensively inactivated by deaminases in plasma and tissues. The plasma half-life of bleomycin is also short, due to renal excretion and absence of protein binding.

Drug Action Relative to the Cell Cycle

Most human cancers have a low growth fraction, the fraction of cells within a tumor which are actively proliferating at any given time. At the same time, most anti-cancer drugs are selectively toxic to dividing cells, or "proliferation-dependent." Prolonged drug infusion will expose more tumor cells to the drug during the sensitive phases of the cell cycle, resulting in greater tumor cell killing. However, toxicity to normal proliferating tissues such as the bone marrow and gastrointestinal tract is also increased by such prolonged infusion.

Drugs have been classified according to their degree of proliferation dependence and specificity for a particular phase of the cell cycle (3). Cytarabine, bleomycin, and the vinca alkaloids are particularly schedule-dependent in experimental systems. The dose-response curves for these drugs in vitro exhibit a characteristic plateau if the duration of drug exposure is held constant, so that increasing drug concentrations do not result in increased cell killing, (4). However, increasing the duration of drug exposure will result in increased cell killing at a given concentration of these agents. Other drugs, notably the alkylating agents, do not exhibit this type of plateau with increasing dose.

Differential Effects on Tumors and Normal Tissues

The threshold of drug concentration required for a given drug to produce cytotoxicity may vary among different tissues and tumors. In some cases, administration of a drug by continuous infusion may reduce toxicity to a normal tissue without loss of antitumor efficacy, by maintaining a lower peak drug concentration than that achieved by bolus injection. Examples of such threshold effects include the reduced bone marrow toxicity of 5-fluorouracil with continuous infusion (5-9), selective reduction of the pulmonary toxicity of bleomycin (10-12), and selective reduction of the cardiac toxicity of doxorubicin (13-20). The biochemical basis for these differences in thresholds of toxicity among various normal tissues and tumors is not well understood.

Drug Resistance

One of the major factors governing the response of human tumors to chemotherapeutic agents is biochemical drug resistance. Generally, manipulating the schedule of drug administration does not overcome such drug resistance, but only increases therapeutic efficacy in tumors which are already at least moderately sensitive to the drug. Biochemical resistance to the currently available agents is prevalent in many types of human tumors, and drug resistant mutations commonly emerge during the course of treatment in tumors which were initially sensitive. Since an alteration in drug scheduling generally will not overcome biochemical drug resistance when present, it has been difficult to demonstrate the superiority of continuous infusion regimens in many clinical trials.

Chemical Properties of Drugs

The stability and solubility of drugs are important considerations in the scheduling of administration. Many alkylating agents are chemically unstable in solution, and are best administered by short infusion or bolus injection. The limited solubility of etoposide requires large volumes for safe administration, and thus etoposide cannot be delivered via pumps which have a limited reservoir volume. Potent vesicants, such as the anthracyclines and vinca alkaloids, require central venous access if administered by continuous infusions, to avoid extravasation and local tissue necrosis with prolonged administration in a peripheral vein.

BLEOMYCIN

Bleomycin is a mixture of glycopeptides which are useful for the treatment of testicular and ovarian germ cell cancers, lymphomas, and squamous carcinomas from various sites. The major dose-limiting toxicity of bleomycin is pulmonary fibrosis. Its mechanism of action involves binding to DNA, the formation of a complex of bleomycin, ferrous ion, and molecular oxygen, and the generation of free radical moieties which result in single and double strand breaks in DNA. The drug is highly proliferation dependent, being most effective during the G2 and M phases of the cell cycle, and also having a short plasma half-life of less than 2 hours.

Continuous infusion of bleomycin has been shown to both increase the therapeutic efficacy and decrease pulmonary toxicity in a mouse model, compared to bolus injections (10,21). Sikic et al. compared the effects of 3 different schedules of administration of bleomycin: 7-day continuous infusion (using the Alzet^R mini-pumps, Alza Corporation, Palo Alto, CA), bolus injections twice weekly (2 doses, s.c.), or bolus injections twice daily for 5 days (10 total doses). Antitumor efficacy was assessed in mice bearing Lewis lung carcinoma, and pulmonary toxicity was measured by total lung hydroxyproline as an index of lung collagen accumulation (10). The results are presented in Table 1:

Table 1. Improved Therapeutic Index of Bleomycin by Continuous Infusion in Mice.

Bleomycin Schedule	Lung Toxicity, % Lung Collagen	Antitumor Effect, % Tumor Shrinkage
Continuous Infusion	110%*	84%*
Bolus x 2	120%	50%
Bolus x 10	131%	68%

* Continuous infusion resulted in significantly less pulmonary toxicity and greater antitumor efficacy than the 2 schedules of bolus injections, P less than 0.05. The total dose was the same in all 3 schedules (10).

The reason for this differential effect of continuous infusion on therapeutic efficacy and pulmonary toxicity is not known. One possibility is that bolus injection results in high pulmonary peak drug levels, which saturate pathways for drug inactivation or repair of drug-induced damage. It is possible that the mechanism for the pulmonary toxicity of bleomycin may differ from its antitumor action, which is thought to result from DNA cleavage. Continuous infusions of bleomycin were also significantly more effective than bolus injections in the experimental therapy of mice bearing P-388 leukemias (21).

There are no randomized, controlled trials in man comparing the efficacy and toxicity of bleomycin administered by continuous infusion vs. bolus injection. However, several clinical studies have reported either increased antitumor efficacy or decreased pulmonary toxicity with bleomycin infusion, compared to historical controls in which the drug was administered by bolus injection (11, 12, 22, 23). At the M.D. Anderson Hospital, before the advent of cisplatin in regimens for testicular carcinomas, the combination of bleomycin and vinblastine resulted in a complete remission rate of 38% when bleomycin was administered by infusion, compared to 15% using the same drugs when bleomycin was given by bolus injection (22). In patients with squamous carcinomas of the cervix, continuous infusions of bleomycin produced a 30% response rate in 32 patients, including two complete remissions, compared to only a 9% response rate for patients treated by bolus injections (11). Pulmonary toxicity did not appear to be increased in this study, although the incidence of alopecia was higher with bolus injection, and cutaneous toxicity (hyperkeratosis and hyperpigmentation) occured in almost all patients.

The Southwest Oncology Group performed a randomized trial in carcinoma of the cervix combining vincristine and mitomycin C with bleomycin, in which three different schedules of bleomycin were compared: weekly and twice-weekly injections, and 4-day infusions. Patients treated with continuous infusions of bleomycin had a significantly longer survival compared to those treated with weekly injections, 6 vs. 4 months (11). No major pulmonary toxicity was observed among the 42 patients treated by continuous infusion, whereas 6 out of 53 patients treated with twice-weekly injections developed severe pulmonary toxicity.

Prospective studies of pulmonary function were performed in a group of 15 patients receiving 7-day continuous infusions of bleomycin, at 15 units/M^2/day (23). None of these patients developed clinically overt pulmonary toxicity, and only two patients showed minor decreases in vital capacity and carbon monoxide diffusing capacity. Further randomized, controlled studies comparing schedules of bleomcyin administration in patients should be done to confirm these findings.

CYTARABINE

Cytarabine (cytosine arabinoside; ara-C) is a pyrimidine anti-metabolite which kills cells during S-phase or DNA synthesis, by direct incorporation of the nucleotide ara-CTP into DNA and by inhibition of DNA polymerase. It is primarily used in the treatment of acute leukemias. The plasma half-life of ara-C is only 10 minutes, with rapid inactivation due to enzymatic deamination. This short half-life and S-phase specificity suggests that continuous infusion therapy or frequent drug injection might increase ara-C efficacy. Table 2 illustrates the beneficial effect of more frequent dose administration of cytarabine in mice bearing L1210 leukemia (24):

Table 2. Schedule Dependence of Cytarabine in Mice with L1210 Leukemia

Dose, mg/kg	Schedule	% Cures
2,500	single injection	0
60	daily x 6	0
15	every 3 hrs. x 8	3
15	every 3 hrs. x 16	58

In clinical studies, continuous intravenous infusions of ara-C for up to 7 days produced no hematologic toxicity at doses below 30 mg/M^2/day (25). Above this dose increasing myelosuppression was noted up to a plateau of 1.5 grams/M^2/day. In contrast, injections of up to 4.2 grams of ara-C as a single intravenous bolus did not result in myelosuppression (26). Nausea and vomiting were greater by bolus injection compared to continuous infusion. Continuous infusion produced greater myelosuppression than daily i.v. bolus injection for 5 days at doses of 50-100 mg/M^2 (27).

Remission rates of 39% were noted in the initial studies of continuous infusion of ara-C at 200 mg/M^2 in acute leukemia (28). Toxicities included moderate to severe leukopenia, nausea, and vomiting. Complete remission rates of 20% and 38% respectively were noted with continuous infusions of 48 and 120 hours of ara-C (29). The total dose of ara-C was similar in these 2 groups, thus confirming observations in pre-clinical models that schedule of administration of ara-C was a more important determinant of efficacy than dose.

More recent studies of ara-C scheduling have utilized the drug in combination with other agents, especially anthracyclines (30). A randomized trial which compared infusions to bolus injections every 12 hours of ara-C in combination with daunorubicin in acute leukemia resulted in higher complete remission rates for patients who received the ara-C by continuous infusion. However, these differences were of borderline statistical significance, and illustrate the problem of studying the schedule of drug administration as a variable in a heterogeneous group of patients where there are also other highly active agents which may obscure the effects of drug scheduling (30).

One possible advantage of continuous infusions of ara-C is the achievement of relatively high drug levels in the cerebro-spinal fluid, approximating one-half of the plasma drug concentration (31, 32). Thus, infusions of ara-C may be useful in treating the types of leukemias and lymphomas which have a high incidence of central nervous system involvement.

Ara-C is not a vesicant or irritating agent. Therefore, it can be safely administered by continuous infusion via a peripheral venous catheter, or even as a continuous subcutaneous infusion.

FLUOROURACIL

Fluorouracil (5-FU) is a pyrimidine anti-metabolite, which inhibits DNA synthesis by binding the enzyme thymidylate synthetase. 5-FU may also be incorporated into RNA as a fraudulent nucleotide. It is used in carcinomas of the breast and gastrointestinal tract. The short plasma half-life of the drug (10 minutes) and its relative specificity for S-phase cells support the administration of 5-FU by continuous infusion.

Continuous infusions of 5-FU are well tolerated at doses of 1.0 to 1.4 grams/day, or 22.5 to 30 mg/kd/day, for up to 5 days (5). Long-term infusions of up to 30 days are well tolerated at doses up to 300 mg/M^2/day (33). The spectrum of toxicities of 5-FU is altered by the schedule of administration. Intravenous bolus 5-FU produces relatively more myelosuppression, while continuous infusions of 5-FU are much less myelosuppressive but result in increased gastrointestinal toxicity, manifested by stomatitis and diarrhea (6-9, 33, 34). The decreased myelosuppression of continuous infusions of 5-FU offers a distinct advantage for this form of administration. However, there has been no convincing evidence of increased antitumor efficacy by such infusions compared to daily bolus injections of the drug (6, 8, 9, 35, 36). Because of the relative lack of myelosuppression of 5-FU by continuous infusion, several clinical studies have combined 5-FU infusion with other cytotoxic but myelosuppressive agents (37-41). The results of these trials were not encouraging, probably because the gastrointestinal cancers in these studies were generally refractory to the agents used. The addition of 5-FU infusion to cyclophosphamide in ovarian carcinomas did yield superior remission rates and survival compared to historical controls with cyclophosphamide alone (42). Another uncontrolled study of infusion 5-FU combined with radiation therapy in patients with esophageal carcinoma claimed superior results compared to historical controls using radiation therapy alone (43).

The major advantage of continuous infusions of 5-FU therefore appears to be a marked reduction in bone marrow toxicity, although there is increased stomatitis and diarrhea by this mode of administration compared to bolus injections. Although 5-FU is not considered a vesicant agent, phlebitis, venous sclerosis, and hyperpigmentation of overlying skin are common after two or three days of infusion through a peripheral vein. Therefore, peripheral venous access sites for 5-FU infusions should be checked daily and changed every two to three days.

DOXORUBICIN

Doxorubicin is an anthracycline compound with a broad spectrum of activity in leukemias, lymphomas, and various solid tumors. Its major acute toxicity is myelosuppression, but the drug also produces cardiac toxicity which is chronic and cumulative and thus limits the total dose which can be administered. This cardiac toxicity is thought to be due to the formation of oxygen-derived free radicals via electron transport at the quinone group of the anthracycline ring. The mechanism of anti-tumor cytotoxicity is probably different, involving intercalation of the drug between base pairs of DNA.

The cardiac toxicity of doxorubicin has been shown to be reduced by divided doses compared to large single doses in various animal models, without compromising therapeutic anti-tumor efficacy (15, 16). Thus, the pre-clinical data suggest that alterations of the schedule of administration may separate out the cardiotoxic effects of doxorubicin from anti-tumor effects.

Clinical studies confirm that the risk for doxorubicin cardiac toxicity is decreased both by continuous infusion therapy (13, 14), and by more frequent administration of lower doses of the drug, weekly rather than every 3 weeks (17-20).

Weekly administration also results in less nausea, vomiting and cardiac toxicity, but more severe stomatitis, than every 3 week therapy. Anti-tumor efficacy did not differ among the various schedules. However it should be emphasized that none of these studies are prospectively randomized comparisons of drug scheduling. Durations of the continuous infusions have ranged from 24 to 96 hours, and the optimal duration of the infusion has not been established.

Doxorubicin is a potent vesicant agent, and therefore delivery by continuous infusion requires a central venous catheter to avoid drug extravasation.

OTHER DRUGS

The vinca alkaloids, **vincristine** and **vinblastine**, are cell cycle phase-specific drugs which bind to tubulin and inhibit microtubule assembly. Their cytotoxic effects are markedly dependent on the duration of drug exposure in cell culture (44). There have been several recent uncontrolled studies of administration of these drugs by continuous infusion, with anecdotal reports of anti-tumor responses in patients who had previously failed conventional bolus injections (1, 2). The toxicities of these various modes of administration appear to be similar. However, the vinca alkaloids are potent vesicants, requiring central venous access for safe administration by prolonged infusion.

Etoposide is a plant-derived epi-podophyllotoxin which produces DNA scission and is S-phase specific. Its poor solubility dictates large volumes of administration, which is impractical for continuous infusion in portable out-patient infusion pumps. Divided doses of etoposide (daily x 3 or 5 days) are superior to single dose therapy both in pre-clinical models and in a randomized trial in patients with small cell lung cancer (1, 2). However, daily bolus injections are probably as effective as continuous infusions for this agent.

Cisplatin is a coordination complex of platinum which produces cross-linking of DNA similar to alkylating agents. Continuous infusions of cisplatin have been utilized in an attempt to reduce the severe nausea and risk for renal toxicity of the drug. However, the renal toxicity is usually not a problem with bolus injection if the patients receive vigorous hydration before, during, and after cisplatin. It is not clear from the uncontrolled studies whether prolongation of the infusion time ameliorates the nausea of the drug. A controlled trial comparing 1 hour to 24 hour infusion of cisplatin has recently been completed by the Gynecologic Oncology Group (Protocol #64 for carcinoma of the cervix), and the results of this study should soon be available.

CONCLUSIONS

There has been increasing interest in the possible advantages of continuous infusion therapy of anti-cancer drugs. The recent improvements in the technology for intravenous access and out-patient infusion pumps have made such treatment more practical and cost-effective. However, the ultimate cost-benefit of such approaches has yet to be determined.

Vesicant agents such as doxorubicin and the vinca alkaloids should not be infused long-term by peripheral vein. Infusion therapy with these drugs should be administered via central venous catherers, to minimize the risk of drug extravasation.

Continuous infusion of certain agents offers the possibility of reducing toxicities while retaining anti-tumor efficacy. Notable examples include the pulmonary toxicity of bleomycin, the cardiac toxicity of doxorubicin, and the bone marrow toxicity of flourouracil. For bleomycin, continuous infusion also appears to increase its anti-tumor efficacy, thus further improving the therapeutic index. The efficacy of cytarabine may also be improved somewhat by continuous infusion compared to bolus injection.

A major problem with much of the clinical literature regarding schedules of anti-cancer drug administration is the lack of rigorous control groups addressing the issue of schedule. In those studies which are appropriately controlled, it has been difficult to prove major clinical benefits in terms of increased anti-tumor efficacy because of alterations of drug scheduling.

An important reason for this difficulty may be the biochemical drug resistance to currently available agents which is prevalent in many human cancers, and which cannot be overcome by an alteration in schedule of drug administration. The marked heterogeneity in tumor drug sensitivity among tumors of any given type makes it difficult to statistically prove superior anti-tumor efficacy in such trials. However, the reduction in risk of certain serious toxicities has been easier to demonstrate, and does provide a substantial rationale for continuous infusion therapies of some of these agents.

REFERENCES

1. Carlson RW and Sikic BI: Continuous infusion or bolus injection in cancer chemotherapy. Ann. Intern. Med. 99: 823, 1983.
2. Vogelzang NJ: Continuous infusion chemotherapy: a critical review. J. Clin. Oncol. 2: 1289, 1984.
3. Bruce WR, Meeker BE, and Valeriote FA: Comparison of the sensitivity of normal hematopoietic and transplanted lymphoma colony-forming cells to chemotherapeutic agents administered in vivo. J. Nat. Cancer Inst. 37: 233, 1966.
4. Skipper HE, Schabel, Jr. RM, and Wilcox WS: Experimental evaluation of potential anticancer agents. XXI. Scheduling of arabinosylcytosine to take advantage of its S-phase specificity against leukemia cells. Cancer Chemother. Rep. 51: 125, 1967.
5. Lemon HM: Reduction of 5-fluorouracil toxicity in man with retention of anticancer effects by prolonged intravenous administration in 5% dextrose. Cancer Chemother. Rep. 8: 97, 1960.
6. Reitemeier RJ and Moertel CG: Comparison of rapid and slow intravenous administration of 5-flurouracil in treating patients with advanced carcinoma of the large intestine. Cancer Chemother. Rep. 25: 87, 1967.
7. Staley CJ, Hart JT, Van Hagen F, and Preston FW: Various methods of administering 5-fluorouracil. Cancer Chemother. Rep. 20: 107, 1962.
8. Seifert P, Baker LH, Reed ML, and Vaitkevicius VK: Comparison of continuously infused 5-fluorouracil with bolus injection in treatment of patients with colorectal adenocarcinoma. Cancer 36: 123, 1975.
9. Hum GJ and Bateman JR: 5-day IV infusion with 5-fluorouracil (5-FU: NSC-19893) for gastroenteric carcinoma after failure on weekly 5-FU therapy. Cancer Chemother. Rep. 59: 1177, 1975.
10. Sikic BI, Collins JM, Mimnaugh, EG, and Gram TI: Improved therapeutic index of bleomycin when administered by continuous infusion in mice, Cancer Treat. Rep. 62: 2011, 1978.
11. Baker LH, Opipari MI, Wilson H, Bottomley R, and Coltman, Jr. CA: Mitomycin C, vincristine, and bleomycin therapy for advanced cervical cancer. Obstet. Gynecol. 52: 146, 1978.
12. Krakoff IH, Cvitkovic E, Currie V, Teh S, and LaMonte C: Clinical pharmacologic and therapeutic studies of bleomycin given by continuous infusion. Cancer 40: 2027, 1977.

13. Legha SS, Benjamin RS, MacKay B, Ewer M, Wallace S, Valdivieso M, Rasmussen SL, Blumenschein GR, and Freireich EJ. Reduction of doxorubicin cardiotoxicity by prolonged continuous intravenous infusion. Ann. Inter. Med. 96: 133, 1982.
14. Legha SS, Benjamin RS, MacKay B, Yap HY, Wallace S, Ewer M, Blumenschein GR and Freireich EJ: Adriamycin therapy by continuous intravenous infusion in patients with metastatic breast cancer. Cancer 49: 1762, 1982.
15. Pacciarini MA, Barbieri B, Colombo T, Broggini M, Grattini S, and Donelli MG: Distribution and antitumor activity of adriamycin given in high-dose and a repeated low-dose schedule to mice. Cancer Treat. Rep. 62: 791, 1978.
16. Solcia E, Ballerini L, Bellini O, Magrini U, Bertazzoli C, Tosana G, Sala L, Balconi F, and Rallo F: Cardiomyopathy of doxorubicin in experimental animals. Factors affecting the severity, distribution and evolution of myocardial lesions. Tumori 67: 461, 1981.
17. Weiss AJ, Metter GE, Fletcher WS, Wilson WL, Grage TB, and Ramirez G: Studies on adriamycin using a weekly regimen demonstrating its clinical effectiveness and lack of cardiac toxicity. Cancer Treat. Rep. 60: 813, 1976.
18. Weiss AJ, and Manthel RW: Experience with the use of adriamycin in combination with other anticancer agents using a weekly schedule, with particular reference to lack of cardiac toxicity. Cancer 40: 2046, 1977.
19. Creech RH, Catalano RB, and Shah MK: An effective low-dose adriamycin regimen as secondary chemotherapy for metastatic breast cancer patients. Cancer 46: 433, 1980.
20. Chlebowski RT, Paroly WS, Pugh RP, Hueser J, Jacobs EM, Pajak TF, and Bateman JR: Adriamycin given as a weekly schedule without a loading course: clinically effective with reduced incidence of cardiotoxicity, Cancer Treat. Rep. 64: 47, 1980.
21. Peng YM, Alberts DS, HSG, Mason N, and Moon TE: Antitumour activity and plasma kinetics of bleomycin by continuous and intermittent administration. Br J. Cancer 41: 644, 1980.
22. Samuels ML, Johnson DE, and Holoye PY: Continous intravenous bleomyicn (NSC-125066) therapy with vinblastine (NSC49842) in stage III testicular neoplasia. Cancer Treat. Rep. 65: 419, 1981.
23. Cooper KR, and Hong WK: Prospective study of the pulmonary toxicity of continuously infused bleomycin. Cancer Treat. Rep. 65: 419, 1981.
24. Skipper HE: Cancer Chemotherapy, Volume 1: Reasons For Success and Failure in Treatment of Murine Leukemias with the Drugs Now Employed in Treating Human Leukemias, University Microfilms International, Ann Arbor, MI, 1978.
25. Ellison RR, Carey RW, and Holland JR: Continuous infusions of arabinosyl cytosine in patients with neoplastic disease, Clin. Pharmacol. Ther. 8: 800, 1967.
26. Frei III, E, Bickers JN, Hewlett JS, Lane M, Lery WV, and Talley RW: Dose, schedule and antitumor studies of arabinosyl cytosine (NSC 63878), Cancer Res. 29: 1325, 1969.
27. Burke PJ, Serpick AA, Carbone PP, and Tarr N: A clinical evaluation of dose and schedule of administration of cytosine arabinoside (NSC 63878), Cancer Res. 28: 274, 1968.
28. Bodey GP, Coltman CA, Hewlett JS and Freireich EJ: Cytosine arabinoside (NSC 63878) therapy for acute leukemia in adults, Cancer Chemother. Rep. 53: 59, 1969.
29. Southwest Oncology Group: Cytarabine for acute leukemia in adults: effect of schedule on therapeutic response, Arch. Intern. Med. 133: 251, 1974.
30. Rai KR, Holland JF, Glidewell OJ, Weinberg V, Brunner K, Obrecht JP, Preisler HD, Nawabi IW, Prger D, Carey RW, Cooper MR, Haurani F, Hutchison JL, Silver RT, Falkson G, Wiernik P, Hoagland HC, Bloomfield CD, James GW, Gottlieb A, Ramanan SV, Blom J, Nissen NI, and Bank A: Treatment of acute myelocytic leukemia: a study by Cancer and Leukemia group B, Blood 58: 1203, 1981.

31. Weinstein HJ, Griffin TW, Feeney J, Cohen HJ, Propper RD, and Sallan SE: Pharmacokinetics of continuous intravenous and subcutaneous infusions of cytosine arabinoside. Blood 59: 1351, 1982.
32. Ho DHW: Potential advances in the clinical use of arabinosyl cytosine. Cancer Treat. Rep. 61: 717, 1977.
33. Lokich J, Bothe Jr, A, Fine N, and Perri J: Phase 1 study of protracted venous infusion of 5-fluorouracil. Cancer 48: 2565, 1981.
34. Sullivan RD, Young CW, Miller E, Glatstein N, Clarkson B, and Burchenal JH: The clinical effects of the continuous administration of fluorinated pyrimidines (5-fluorouracil and 5-fluoro-2'-deoxyuridine), Cancer Chemother. Rep. 8: 77, 1960.
35. Moertel CG, Schutt AJ, Reitemeier RJ, and Hahn RG: A comparison of 5-fluorouracil administered by slow infusion and rapid injection. Cancer Res. 32: 2717, 1972.
36. Hartman Jr., HA, Kessinger A, Lemon HM, and Foley JF: Five-day continuous infusion of 5-fluorouracil for advanced colorectal, gastric, and pancreatic adenocarcinoma. J. Surg. Oncol. 11: 227, 1979.
37. Greco FA, Richardson RL, Schulman SF, and Oldham RK: Combination of constant-infusion 5-fluorouracil, methyl-CCNU, mitomycin C, and vincristine in advanced colorectal carcinoma. Cancer Treat. Rep. 62: 1407, 1978.
38. Kane RC, Cashdollar MR, and Bernath AM: Treatment of advanced colorectal cancer with methyl-CCNU plus 5-day fluorouracil infusion. Cancer Treat. Rep. 62: 1521, 1978.
39. Bedikian AY, Stab R, Livingston R, Valdivieso M, Burgess MA, and Bodey GP: Chemotherapy for colorectal cancer with 5-fluorouracil, cyclophosphamide, and CCNU: comparison of oral and continuous IV administration of 5-fluourouracil. Cancer Treat. Rep. 62: 1603, 1978.
40. Buroker T, Kim PN, Gropp C, McCracken J, O'Bryan R, Panettiere F, Cottman C, Bottomley R, Wilson H, Bonnet J, Thigpen T, Vaitkevicius VK, Hoogstraten B, Heilbrun L: 5-FU infusion with mitomycin-C versus 5-FU infusion with methyl-CCNU in the treatment of advanced colon cancer. Cancer 42: 1228, 1978.
41. Buroker T, Kim PN, Groppe C, McCracken J, O'Bryan R, Panettiere F, Costanzi J, Bottomley R, King GW, Bonnet J, Thigpen T, Whitecar J, Hass C, Vaitkevicius VK, Hoogstraten B, and Heilbrun L: 5-FU infusion with mitomycin-C versus 5-FU infusion with methyl-CCNU in the treatment of advanced upper gastrointestinal cancer. Cancer 44: 1215, 1979.
42. Izbicki RM, Baker LH, Samson MK, McDonald B, and Vaitkevicius VK: 5-FU infusion and cyclophosphamide in the treatment of advanced ovarian cancer. Cancer Treat. Rep. 61: 1573, 1977.
43. Byfield JE, Barone R, Mendelsohn J, Byfield JE, Barone R, Mendelsohn J, Frankel S, Quinol L, Sharp T, and Seagren S: Infusional 5-fluorouracil and x-ray therapy for non-resectable esophageal cancer. Cancer 45: 703, 1980.
44. Jackson, Jr., DV, and Bender RA: Cytotoxic thresholds of vincristine in a murine and a human leukemia cell line in vivo. Cancer Res. 39: 4346, 1979.

PHARMOCOLOGY AND THERAPEUTIC EFFICACY OF BLEOMYCIN ADMINISTERED BY CONTINUOUS INFUSION

Irwin H. Krakoff
M. D. Anderson Hospital and Tumor Insitute
Division of Medicine
6723 Bertner Avenue
Houston, Texas 77030

Bleomycin was isolated by Umezawa[1]. It has been in clinical use for more than 15 years. The initial clinical studies performed by Ichikawa, et al[2] demonstrated useful activity in the squamous carcinomas. Subsequent studies have demonstrated useful therapeutic activity in lymphomas and germ cell tumors[3-5].

The preparation originally developed for clinical trial is a mixture of polypeptides. Although this mixture can be separated into several discrete fractions none of those has been shown to be more effective or less toxic than the mixture and the current commercially available material (Blenoxane) is still composed of the mixture as originally studied.

The initial clinical studies were done with bleomycin given intraveneously at doses of 15 or 30 milligrams once or twice weekly. In subsequent studies, generally similar dose regimens were used. When given by daily injection, toxicity appeared to be no different from that which occurred with less frequent administration; mucocutaneous toxicity appeared to be related to the total cumulative dose given and the incidence of pulmonary fibrosis was reported to be similar with daily and weekly injections. Therapeutic activity of bleomycin given daily appeared to be better in lymphomas than had been achieved using weekly or semi-weekly schedules, although controlled studies were not done.

Bleomycin was demonstrated[6] to be cell-cycle specific inhibiting cell progression at the S/G_2 boundary and in the mitotic period. Cells in other stages of the cell cycle were much less susceptible to inhibition by bleomycin. It could be concluded, therefore, that for bleomycin, as for other highly cell cycle specific agents, it was important for the drug to be present throughout the cell cycle.

Pharmacokinetic studies[7] revealed a half life in the serum of less than 2 hours following intravenous injection. It was concluded that on the basis of the very short retention time in the serum and the requirement for nearly continous exposure that intermittent injection would be less likely to produce optimal effects than would the maintenance of effective blood levels for a longer period of time, by means of continuous intravenous infusion.

In order to test this hypothesis, a study was done at Memorial Hospital[8] in which bleomycin was given by continuous intravenous infusion to patients with a variety of far advanced unresectable malignant neoplastic diseases. Because of the potential for greatly increased toxicity, this reevaluation was approached as a Phase I study with administration initially at much smaller doses than usual, i.e. 0.02mg/kg/day, approximately 1/10 the dose that had been previously studied. There was no acute toxicity of that dose and the daily dose was therefore gradually escalated to a level of .025mg/kg daily (approximately 10mg/m^2).

Bleomycin blood levels were measured by bioassay*. However, at the doses used, blood levels could be measured only in the few hours immediately after injection. Therefore, ^{111}Indium-bleomycin was mixed with commercial bleomycin to provide a dose of 75 microcuries of ^{111}Indium/kg (5 millicuries for an average adult in a dose of 0.25mg/kg of bleomycin). The methodology for preparation of the material and its analysis in body fluids has been described in a previous publication[8].

Pulmonary function was studied by measurement of total lung capacity and carbon monoxide diffusing capacity.

In these studies, as in earlier ones in which bleomycin had been given by bolus intravenous injection, most manifestations of toxicity were related to total cumulative dose. Mucocutaneous toxicity occurred regularly between the 7th and 11th day of the infusion at a dose of 0.25mg/kg per day. Since, in that study therapy was continued until the development of limiting toxicity, mucocutaneous toxicity occurred in nearly 100% of the patients. It was characterized by painful erythema and ulceration of the tongue and oral mucosa; erythema and superficial ulceration in areas of skin subjected to pressure was also very common. Alopecia occurred in nearly three-quarters of the patients; this was more frequent than had been seen in patients treated with bolus intravenous injections. Fever and minor myelosuppression occurred in a small proportion of patients. The major toxicologic problem was pulmonary insufficiency. Although only six patients in this series developed overt pulmonary toxicity, in five of those it progressed to fatal pulmonary fibrosis. That incidence was not significantly different from what had been seen in our own experience and others with bleomycin given by bolus injection.

---

* Courtesy of Dr. Karl Agre, Bristol Laboratories

Detailed pulmonary function studies were done in the first 35 patients treated in this series; nearly all of those developed significant decrease in total lung capacity and in carbon monoxide diffusion capacity. Neither of those was related to the total dose of bleomycin. It was of some concern that each of the six patients with overt clinical pulmonary toxicity demonstrated marked abrupted reduction in pulmonary function which occurred more than 45 days after the end of the infusion of bleomycin. Thus, although serial monitoring provided evidence of decreased pulmonary function in nearly every patient monitored, it was not valuable in predicting or preventing the occurrence of severe pulmonary toxicity.

In another study reported in 1981, Cooper and Hong[9] described the administration of bleomycin given as a continuous IV infusion at a dose rate very similar to that reported in our previous study. Their patients were followed very carefully with measurements of lung volume, carbon monoxide diffusing capacity, spirometery, arterial blood gases and chest radiographs. They observed no changes in pulmonary function in any of the patients. Additionally none of those patients experienced overt clinical pulmonary toxicity due to bleomycin. It should be noted that none of the patients in that series had previously received therapeutic radiation, chemotherapy or any known pulmonary toxic agents. This is in contrast to patients in the previously reported study in which most patients had received prior chemotherapy and many had received radiotherapy; that difference may be responsible for the difference in pulmonary functional measurements between the two series. Bleomycin given by continuous intravenous infusion has been a component of intensive combination chemotherapy in two studies[10-11] of the treatment of advanced metastatic germ cell tumors of the testis. In both of those, pulmonary toxicity was minimal.

A similar lack of pulmonary toxicity was demonstrated in mice given bleomycin by continuous infusion in a study reported by Sikic et al[12]. In those experiments, bleomycin given twice daily produced a significantly better therapeutic effect than when the same total dose was given twice weekly. Both of those schedules produced pulmonary toxicity. When the same total dose of bleomycin was given by continuous intravenous infusion for 7 days, better inhibition of tumor growth was produced than by either previous schedule. In addition, continuous infusion did not produce pulmonary toxicity as measured by lung hydroxyproline content, a reliable index of tissue collagen levels. Sikic pointed out that multiple injections at either high or low doses produced peak levels of bleomycin, whereas continuous infusion produces a lower continuous blood level. It was postulated that the peak levels might be responsible for pulmonary toxicity and that continuous low levels are less toxic than intermittent high peaks.

In the clinical studies[8] performed at Memorial Hospital several years earlier, bleomycin levels were measured by bioassay. Due to the relative insensitivity of that method, it was not possible to determine whether significant levels remained in the blood for prolonged periods after the serum concentration had fallen to levels below those which were detectable. Serum concentrations of bleomycin could not be measured at all using that method when the drug was given by intravenous infusion. Therefore, we studied blood levels and disappearance curves by following the fate and disappearance of radioactivity after the administration of ^{111}Indium-labeled bleomycin. It appeared that a fair approximation of the biological levels could be made in this way. In a number of experiments it was concluded that ^{111}Indium and bleomycin remained in close association <u>in vitro</u> and <u>in vivo</u>. Using ^{111}Indium measurements as an indication of bleomycin level it was demonstrated that single bolus intravenous doses of bleomycin produced high peak levels falling off rapidly and that continuous intravenous infusion produced much lower levels which were maintained for a much longer period of time. Intramuscular administration of bleomycin produced peak levels which were lower than when the same dose was given intravenously; however, the duration and disappearance of bleomycin from the serum were similar to those seen when the drug was given intravenously. Comparison of aqueous and oil-suspended bleomycin given intramuscularly revealed no difference in peak levels or in persistence of bleomycin in the serum.

In the Memorial Hospital study, therapeutic activity in cervix cancer appeared to be significantly better (i.e. 6% CR, 25% PR) than in earlier studies by the same group of investigators. Otherwise, response rates in that series were similar to those seen in similar neoplasms by a number of investigators. To date, controlled studies have not been done comparing bleomycin given by conventional intravenous bolus injection and continuous infusion. However, <u>in vitro</u> and animal studies and our own clinical pharmacologic studies support the logic of continuous intravenous administration in the effort to decrease pulmonary toxicity and to improve therapeutic effect.

REFERENCES

1. Umezawa, H., Maeda, K., Takeuchi, T., et al: New antibiotics bleomycin A and B. J. Antibiot. (A) 19:200-209, 1966.

2. Ichikawa, T., Nakano, I., and Hirokawa, I.: Bleomycin treatment of the tumors of penis and scrotum. J Urol. 102:699-707, 1969.

3. Yagoda, A., Mukherji, B., Young, C., Etcubanas, E., LaMonte, C., Smith, J., Tan, C., and Krakoff, I. H.: Bleomycin, an antitumor antibiotic - Clinical experience in 274 patients. Ann. Intern. Med. 77:861-870, 1972.

4. Bonadonna, G., DeLena, M., Monfardini, S., Bartoli, C., Bajetta, E., Beretta, G., and Fossati-Bellani, F.: Clinical trials with bleomycin in lymphomas and in solid tumors. Eur. J. Cancer 8:205-215, 1972.

5. Blum, R. H., Carter, S. K., and Agre, K.: A clinical review of bleomycin - A new antineoplastic agent. Cancer 31:903-914, 1973.

6. Barranco, S. C., Luce, J. K., Romsdahl, M. M., and Humphrey, R. M.: Bleomycin as a possible synchronizing agent for human tumor cells in vivo. Cancer Research 33:882-887, 1973.

7. Fujita, H., and Kumura, K.: Blood level, tissue distribution, excretion and activation of bleomycin. Proc. 6th International Congress Chemotherapy 2:25, 1970.

8. Krakoff, I. H., Cvitkovic, E., Currie, V., Yeh, S., and LaMonte, C.: Clinical pharmacologic and therapeutic studies of bleomycin given by continuous infusion. Cancer 40:2027-2037, 1977.

9. Cooper, K. R., and Hong, W. K.: Prospective study of the pulmonary toxicity of continuously infused bleomycin. Cancer Treatment Reports 65:419-425, 1981.

10. Vugrin, D., Herr, H. W., Whitmore, W. F., et al: VAB-6 combination chemotherapy in disseminated cancer of the testis. Ann Intern Med 95:59-61, 1981.

11. Logothetis, C. J., Samuels, M. L., Selig, D., Swanson, D., Johnson D. E., von Eschenbach A. C.: Improved survival with cyclic chemotherapy for nonseminomatous germ cell tumors of the testis. Journal of Clinical Oncology 3:326-335.

12. Sikic, B. I., Collins, J. M., Mimnaugh, E. G., and Gram, T. E.: Improved therapeutic index of bleomycin when administered by continuous infusion in mice. Cancer Treatment Reports 62:2011-2017, 1978.

CONTINUOUS-INFUSION ADRIAMYCIN

Robert S. Benjamin, Sant P. Chawla, Gabriel N. Hortobagyi, Michael S. Ewer, Bruce Mackay, Sewa S. Legha, C. Huberto Carrasco, and Sidney Wallace

Departments of Medical Oncology, Internal Medicine Pathology, and Diagnostic Radiology, The University of Texas System, M.D. Anderson Hospital and Tumor Institute at Houston, Houston, Texas 77030

INTRODUCTION

The use of adriamycin is frequently limited by the development of a cumulative-dose-dependent cardiomyopathy, which increases substantially in incidence at doses ≥ 550 mg/m^2. To avoid the potential development of fatal refractory congestive heart failure (CHF), dose limitation to 450-550 mg/m^2 has been recommended by a variety of authors.[1-3] Since an oncologist may not wish to stop treatment of an individual patient at an arbitrary level, a considerable investment of time and resources has been made in an effort to extend the use of adriamycin or develop analogs with diminished cardiotoxicity to accomplish the same aim.

With adriamycin, two approaches have been taken: individualization of therapy through improved cardiac monitoring, and schedule manipulation. The most important step in the investigation of anthracycline cardiotoxicity was the development of a reliable, quantitative, monitoring system to cause CHF. Thus, the discovery by Dr. Margaret Billingham of specific, cumulative-dose-related, pathologic changes, identifiable and quantifiable by electron-microscopic evaluation endomyocardial biopsies, provided the critical tool for our evaluation.[4,5] After a modification of the grading system by Dr. Bruce Mackay at our institution to allow easier quantitation and reproducibility,[6] we have used endomyocardial biopsy to evaluate the schedule dependency of adriamycin cardiotoxicity.[7,8] We have also studied the value of endomyocardial biopsy as a predictive test for the development of CHF at doses >500 mg/m^2 and correlated biopsy grades with ejection fractions determined by radionuclide cineangiography.[9,10]

Endomyocardial biopsy is the most sensitive indicator of anthracycline cardiotoxicity, frequently showing quantitative changes at doses causing no functional abnormalities.[5,9] Also, biopsy grade can be related in a linear fashion to cumulative dose.[5,9] It is thus the best method to compare treatment schedules and is essential in the evaluation of a new analogue where cumulative-dose guidelines do not exist. Since the average biopsy grade at 400-500 mg/m^2 in the Stanford series was 1.8,[5] we decided to treat patients to a grade 2 biopsy or CHF rather than a fixed cumulative dose.

Dr. Sant Chawla, analyzed the data on over 250 patients treated above 500 mg/m^2 with adriamycin on a variety of schedules. Cardiac biopsies were usually performed every four courses with scans every two courses.[10] There were approximately 400 cardiac biopsies and 1,000 ejection fraction determinations. Twenty-two patients developed CHF; however, only 2 cases were fatal. The risk of developing congestive heart failure within four courses of therapy was strikingly correlated with cardiac biopsy grade, increasing from 4% in 175 patients with low-grade (≤ 1) biopsies to 24% in 49 patients with high grade (≥ 1.5) biopsies. No test is absolute, however, and even with minor changes on cardiac biopsy, there was a 2% incidence of congestive heart failure which we could not attribute to any other factor, although other factors may, in fact, have been involved. There was a 27% incidence of heart failure with a biopsy grade of 1.5 but only a 12% incidence of CHF with a biopsy grade of 2. The reason is that before this analysis was performed, patients received up to four courses of additional adriamycin with a grade 1.5 biopsy but none if they had a grade 2 biopsy.

When we analyzed the cardiac ejection fractions, we found that if the ejection fraction was >65 (the mean baseline at our institution is 72), there was <5% incidence of heart failure within 2 courses of therapy.[10] If the ejection fraction was <50, the incidence of heart failure increased to about 30%, but between 50 and 64, analyzed in 5 point groups, there was no difference, and the risk of CHF was 12-14%. We looked at fall in ejection fraction, which had been reported to be of prognostic value,[11] and saw only a slight suggestion that a >15 point fall was important. When we examined the 15 point fall in ejection fraction more carefully, we found that it wasn't the degree of the fall which was important but rather the value of the final ejection fraction: if it was >50, the incidence of developing heart failure was low, and if it was <50, there was a substantial risk of developing heart failure (Table 1).

Table 1: Risks of Developing Congestive Heart Failure by Fall in Ejection Fraction and Last Ejection Fraction

	< 15 Percent Fall in Ejection Fraction	> 15 Percent Fall in Ejection Fraction	
Number of Patients with Ejection Fraction $\geq$ 50	177	18	
Percent with Heart Failure	7	11	p = 0.57
Number of Patients with Ejection Fraction < 50	2	7	
Percent with Heart Failure	100	29	p = 0.17

Our current monitoring approach is best illustrated by the data in Table 2. Patients with ejection fractions >65 do not routinely have biopsies. Although we can pick up a few extra cases of heart failure by performing the biopsies, the yield is too low to justify the risk involved. Conversely, with an ejection fraction < 50 we no longer perform biopsies because the risks of continuing adriamycin are too high, regardless of the biopsy grade. But in the ejection fraction range of 50-64 we require endomyocardial biopsy since, if the biopsy is a low-grade, there is < 5% risk of heart failure compared with a 33% incidence of patients with high-grade biopsies.[10]

Table 2: Risk of Developing Congestive Heart Failure by Ejection Fraction and Cardiac Biopsy

	Ejection Fraction in Percent		
	>65	50-64	< 50
Biopsy Grade < 1			
Number of Patients	118	51	6
Percent with Heart Failure	2	4	50
Biopsy Grade > 1.5			
Number of Patients	16	30	3
Percent with Heart Failure	13	33	3

Dr. Arthur Weiss reported several years ago that weekly adriamycin administration is less cardiotoxic than standard, every-3-week administration.[12] This observation has been confirmed in another clinical study using heart failure as an end point.[13] Many oncologists were not convinced of the validity of the data, however, until a retrospective analysis of over 4000 patients confirmed the advantage of the weekly schedule with statistical significance.[14] Subsequently, we confirmed the observation with endomyocardial biopsy in only 28 patients, without producing CHF or even requiring high cumulative doses of adriamycin.[8] A similar observation was made by Torti et al., who estimated that cumulative weekly doses could be increased about 200 mg/m^2 over those of standard, every-3-week administration for equivalent biopsy grades.[15]

Why should weekly adriamycin be less toxic? The only reasonable explanation is that there is a lower peak drug level, and there is time for the peak to be eliminated before the next dose. The best way to limit peak drug levels is to give a continuous infusion because the peak is inversely proportional to the time over which drug is given. We started a study, therefore, using adriamycin as a continuous infusion at a fixed dose of 60 mg/m^2 as a single agent in patients with CMF-resistant breast cancer.[16] All patients were required to have central venous catheters. We used percutaneously placed Centrasil or Intrasil catheters, which were left in place as long as they were needed. The drug was usually administered over 24 hours. If the 24-hour infusion was tolerated, we escalated the duration of infusion 100% and kept the dose the same, going from 24 to 48 to 96 hours. Peak levels were lowered about 13-fold. In that study, there was a 50% response rate among 26 evaluable patients. We did not suggest that this was a higher response rate than might have been seen with standard adriamycin. Our primary concern was that we might eliminate the toxicity and all the antitumor effects as well, but the substantial antitumor effects observed suggested a selective reduction in toxicity. Myelosuppression was basically the same as would be seen at the same doses of adriamycin given as a rapid infusion; however, nausea and vomiting were markedly reduced. Only 15% of patients had even moderate nausea and vomiting, primarily with the 24- and 48-hour infusions. In fact, the 96-hour infusion produced essentially no vomiting and very little nausea. In contrast, we did see an increased incidence of mucositis as expected from earlier studies with a daily x 3 schedule.[17]

Our next study used the 96-hour continuous infusion of adriamycin in combination chemotherapy for front-line therapy of sarcomas in the CyADIC regimen which has been used without continuous infusion for the last 10 years.[18] Both adriamycin and DTIC were given by 96-hour infusion at total doses of 60 mg/m^2 and 600 g/m^2, respectively. The two drugs are compatible and stable, and we mixed them together in the same pump. Cyclophosphamide, 600 mg/m^2, was given as a 4-hour infusion on day one. We increased the adriamycin and cyclophosphamide doses by 25% per course up to the highest dose which did not cause infection or severe stomatitis, usually the second level. There was a 53% response rate in the initial patients with soft-tissue sarcomas. A major question was whether we would decrease the response rate by eliminating the peak since sarcomas have a very steep dose-response curve to adriamycin (even though this is a combination, adriamycin is the primary agent). The answer was "no."

The final clinical trial was a study of the FAC regimen in patients with previously untreated, metastatic breast cancer by Dr. Gabriel Hortobagyi.[19] There were two sequential studies: the first with standard-infusion adriamycin, and the second with continuous 48-96 hour infusion. The second dose of 5FU was given on day 5, at the end of the 96-hour adriamycin infusion, an insignificant change from the day 8, used in standard FAC. The treatments were otherwise identical, including the intensification therapy, the concomitant hormonal therapy for ER-positive or -unknown patients, etc. With standard FAC we stopped adriamycin at 450 mg/m^2, going on to CMF. In the continuous-infusion FAC we continued adriamycin until the patient had disease progression or developed a grade 2 biopsy of CHF. The two groups of patients were balanced in terms of important prognostic factors. Very obvious was a change in the acute toxicity pattern. In the 96-hour infusion FAC there was only a 6% incidence of moderately severe nausea and vomiting, even in association with Cytoxan and 5FU, whereas in standard FAC, there was a 46% incidence of moderately severe nausea and vomiting. Conversely, there was an increase in stomatitis, from 5% with standard FAC up to 35% with 96-hour infusion.

If stomatitis becomes dose-limiting, my practice is to go back to a 48-hour infusion, which had about a 15% incidence of both moderately severe nausea and vomiting and stomatitis. Response rates were the same both with regard to complete and complete plus partial remission. The subgroup which benefited most from continuation of adriamycin was the complete responders, whose median time to progression was extended by approximately 10 months with the continuous infusion adriamycin, a time equivalent to the duration of extra adriamycin therapy. Similarly, survival of the complete responders appears to be improved. These differences are not yet statistically significant, however.

We looked at cardiotoxicity with endomyocardial biopsies. Our initial data showing decreased cardiotoxocity with 96-hour continuous infusion compared with rapid infusion have already been published.[7] At each cumulative dose level, the cardiac biopsy grades were lower with the continuous infusion. Table 3 summarizes our data from a group of patients without additional cardiac risk factors who were treated with various infusion durations. The rapid-infusion patients were all screened at 450 mg/m^2 and found to have cardiac biopsy grades $\leq$ 1.5, and, therefore, we thought that they could continue to receive adriamycin. In that group, however, we were able to reach a median cumulative dose of only 550 mg/m^2 and a maximum of 650 mg/m^2. The mean cardiac biopsy grade for these standard patients is 1.3, and 45% of patients had high-grade biopsies. By giving a 24-hour infusion, we produced the same degree of cardiotoxicity at a median cumulative dose of 860 mg/m^2. We were able to give an additional 300 mg/m^2 of adriamycin, but a substantial number of patients had high-grade biopsies or low ejection fractions. With the 96-hour infusions, we were able to increase the cumulative dose up to a median dose of 945 mg/m^2. Nonetheless, the mean biopsy grade was < 1, and < 20% of patients had high-grade biopsies. Thus, we were able to almost double the cumulative dose and still had decreased cardiotoxicity. The reasons that the 96-hour infusion was given in higher doses than the 48-hour infusion is that these were not simultaneous studies and more sarcoma patients. who get the highest dose per course, received 96-hour infusions. When compared at similar dose levels, 96-hour infusions were always slightly less toxic than 48-hour infusions. Similar data were obtained on patients who had cardiac risk factors: 30 patients were treated by 96-hour infusion, and even in the presence of other cardiac risk factors, achieved a mean cumulative dose of 810 mg/m^2 with only 10% high-grade biopsies.

Table 3: Adriamycin Cardiotoxicity by Schedule in Patients Without Risk Factors

	Schedule (Infusion Duration)			
	Rapid	24-hr	48-hr	96-hr
Number of Patients	11	16	16	48
Adriamycin Dose (mg/m^2)				
Median	550	860	700	945
Range	500–650	520–1080	530–920	600–1905
Highest Biopsy Grade				
Mean	1.3	1.3	0.8	0.9
Percent > 1.5	45	40	6	19
Percent with EF < 50	10	31	7	2
Percent with Heart Failure	18	13	0	4

With 96-hour infusion, the median cumulative dose which is required to produce a biopsy grade of 1 is 800 mg/m^2. At 800-1000 mg/m^2, the mean biopsy grade is 0.8, and at 1200 mg/m^2 the mean biopsy grade is 1.0. The highest dose we've given is 1905 mg/m^2.

Patients do get heart failure, even with continuous infusion adriamycin, but contrary to what is usually stated in articles or books,[2-3] the heart failure usually responds to standard therapy.[20] It was fatal in only two of our patients. Heart failure, when picked up by cardiac monitoring, is easily treatable and is actually reversible in some cases. With standard infusion, even with monitoring, at least the way we did it before Dr. Chawla's analysis,[10] the incidence of congestive heart failure increases steeply above 550 mg/m^2. In contrast, we have treated more than 50 patients with 800 mg/m^2 the incidence of CHF increases to about 5% and at 1200 mg/m^2 it increases further to about 10%.

So in summary, we have demonstrated diminished cardiotoxicity as well as diminished nausea and vomiting with continuous infusions of adriamycin, particularly with infusions of 48 hours or longer, and best with 96-hour infusions, the longest duration that we have studied systematically. At least in breast cancer, we're beginning to get data that more adriamycin is better, but it's better only for a selected subgroup of patients: those with complete remission. The diminished cardiotoxicity makes the use of adriamycin more attractive in the adjuvant situation, where increased safety will decrease the chances of long-term complications and make retreatment easy for cured patients who develop second malignancies. At least one alternative would be not to use so much adriamycin in primary treatment, but to treat the majority of patients to a conservative cumulative dose limit of 450-600 mg/m^2 by 96-hour infusion, continuing only in selected patients showing complete remission or continued response and giving just as much or more a second time around for the group of patients you may wish to retreat. I think the question of treatment duration depends on whether your primary goal is palliative or curative, but there's no question that you can deliver twice as much adriamycin by 96-hour infusion with less cardiotoxicity than with a standard amount of adriamycin by standard, every-3-week administration. We recommend a conservative dose limit of 800 mg/m^2 for the 96-hour infusion without the necessity of cardiac monitoring for patients without underlying risk factors.

REFERENCES

1. Minow, R.A., Benjamin, R.S., Lee, E.T., Gottlieb, J.A.: Adriamycin cardiomyopathy: risk factor. Cancer 39:1397-1402.
2. Lefrak, E.A., Pitha, J., Rosenheim, S., Gottlieb, J.A.: A clinicopathologic analysis of adriamycin cardiotoxicity. Cancer 32:302-314, 1973.
3. Blum, R.H., Carter, S.K.: Adriamycin: A new anticancer drug with significant clinical activity. Ann Intern Med 80:249-259, 1974.
4. Billingham, M., Bristow, M.R., Glatstein, E., et al: Adriamycin cardiotoxicity: endomyocardial biopsy evidence of enhancement by irradiation. Am J Surg Path 1:17-23, 1977.
5. Bristow, M.R., Mason, J.W., Billingham, M.E., et al: Doxorubicin cardiomyopathy: evaluation by phonocardiography, endomyocardical biopsy and cardiac catheterization. Ann Intern Med 88:169-175, 1978.
6. Mackay, B., Keyes, L.M., Benjamin, R.S., et al: Cardiac biopsy. Texas Society for Electron Microscopy Journal 11:7-15, 1981.
7. Legha, S.S., Benjamin, R.S., Mackay, B., et al: Reduction of infusion. Ann Intern Med 96:133-139, 1982.

8. Valdivieso, M., Burgess, M.A., Ewer, M.S., et al: Increased therapeutic index of weekly doxorubicin in the therapy of non-small cell lung cancer. A prospective, randomized study. J Clin Oncology 2:207-214, 1984.
9. Ever, M.S., Ali, M.K., Mackay, B., et al: A comparison of cardiac adriamycin. J Clin Oncology 2:112-117, 1984.
10. Chawla, S.P., Benjamin, R.S., Legha, S.S., et al: Role of cardiac biopsy and radionuclide scan in monitoring of adriamycin-induced cardiotoxicity in "Proceedings of the 13th International Congress of Chemotherapy," K.H. Spitzy and K. Karrer, eds., Verlag H. Egermann, Vienna, 1983, pp 490-492.
11. Alexander, J., Dainiak, N., Berger, H.J., et al: Serial assessment of angiocardiography. N Engl J Med 300:278-283, 1979.
12. Weiss, A.J., Manthel, R.W.: Experience with the use of adriamycin in combination with other anticancer agents using a weekly schedule with particular reference to lack of cardiac toxicity. Cancer 40: 2046-2052, 1977.
13. Chlebowski, R.T., Paroly, W.S., Pugh, R.P., et al: Adriamycin given as weekly schedule without a loading course: clinically effective with a reduced incidence of cardiotoxicity. Cancer Treat Rep 64: 47-51, 1980.
14. Von Hoff, D.D., Layard, M.W., Basas, P., et al: Risk factors for doxorubicin-induced congestive heart failure. Ann Intern Med 91: 710-717, 1979.
15. Torti, F.M., Bristow, M.R., Howes, A.E., et al: Reduced cardiotoxicity of doxorubicin delivered on a weekly schedule. Assessment by endomyocardial biopsy. Ann Intern Med 99:745-749, 1983.
16. Legha, S.S., Benjamin, R.S., Mackay, B., et al: Adriamycin therapy by continuous intravenous infusion in patients with metastatic breast cancer. Cancer 49:1763-1766, 1982.
17. Benjamin, R.S.: A practical approach to adriamycin toxicology. Cancer Chemotherap Rep 6:191-194, 1975.
18. Benjamin, R.S. and Yap, B.S.: Infusion chemotherapy for soft tissue sarcomas. IN: "Soft Tissue Sarcomas," Baker LH, ed., Martinus Nijhoff, The Netherlands, 1983, pp 345-350.
19. Hortobagyi, G., Frye, D., Blumenschein, G., et al: FAC with adriamycin by continuous infusion for treatment of advanced breast cancer. Proc Am Soc Clin Oncol 2:105, 1983.
20. Haq, M.M., Legha, S.S., Choksi, J., et al: Doxorubicin-induced congestive heart failure in adults. Cancer 56:1361-1365, 1985.

5-FLUOROURACIL PLUS THYMIDINE OR LEUCOVORIN BY CONTINUOUS I.V. INFUSION IN THE TREATMENT OF ADVANCED COLORECTAL CARCINOMA

Fabio Trave and Youcef M. Rustum

Grace Cancer Drug Center, Roswell Park Memorial Institute

Buffalo, NY 14263

INTRODUCTION

5-Fluorouracil (FUra) has been the drug of choice in the treatment of colorectal cancers (1,2), and it is widely utilized in a variety of other malignancies (3). FUra exerts its antiproliferative effect following metabolic activation to various nucleotides (Chart 1). 5-Fluorouridine triphosphate (FUTP), due to its resemblance with uridine triphosphate (UTP), is incorporated into RNA (4-7); the consequence of this incorporation is the production of fraudulent mRNA, rRNA and tRNA which can ultimately cause cell death. The other proposed mechanism for FUra cytotoxicity is the inhibition of thymidylate synthetase (dTMP-S) by 5-fluorodeoxyuridine monophosphate (FdUMP) (8-11), leading to decreased thymidine triphosphate (dTTP) pools and to inhibition of DNA synthesis. The biochemical mechanism by which FdUMP binds to the dTMP-S involves a cofactor, $N^{5,10}$methylene tetrahydrofolic acid ($N^{5,10}CH_2FH_4$). Santi et al. have calculated that the dissociation constant (Kd) of the ternary complex FdUMP-dTMP-S$N^{5,10}CH_2FH_4$ is of the order of 5×10^{-11}M. In absence of the reduced folate cofactor, FdUMP binding to dTMP-S is relatively weak, with a Kd of about 10^{-5}M (9). An additional proposed mechanism of cytotoxicity of FUra is its incorporation into DNA (12-15). To date, little is known about the biological significance of this finding.

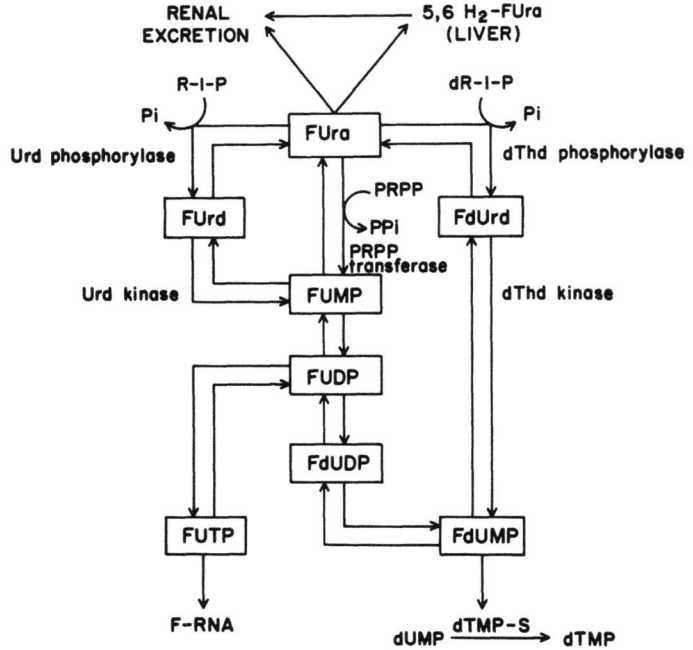

Chart 1. Metabolism of FUra *In Vivo*.

Pharmacological (narrow therapeutic index, short elimination half-life) and cytokinetic (phase-specificity) rationales were proposed to suggest the need for an evaluation of this agent by continuous i.v. infusion. However, clinical trials over the last fifteen years have demonstrated that different FUra regimens of i.v. administration, although appearing to affect the spectrum and degree of host toxicity (Table 1), do not seem to alter the therapeutic efficacy of this agent to any significant extent. In fact, despite the variety of the schedules employed, the response rate of colorectal cancer patients to this agent is still about 20% (1,3).

One approach to increase the selectivity of antimetabolites has been the administration of per se non-cytotoxic substances which can 'force' the metabolism of a drug in a selective way, resulting in an improved therapeutic index (16). An appropriated modulation with metabolites or cofactors could either increase the sensitivity of tumor cells or reduce the sensitivity of normal cells to a given anticancer drug, resulting in a favourable net effect. The clinical use of high-dose methotrexate with leucovorin rescue is a practical example of such useful modulations.

Table 1. Toxicity of FUra : I.V. Push vs. I.V. Infusion

Study	No. of Pts and Tumor Type	Doses and Schedules	B.M. Toxicity (1)	G.I. Toxicity (2)
Ref. 30	149 pts. Colorectal CA	13.5mg/kg I.V. Push 20-25mg/kg 2hr Inf.	63% 39%	50% 33%
Ref. 31	70 pts. Colorectal CA	12mg/kg I.V. Push 30mg/kg 52 days Inf.	72% 12%	8% 65%
Ref. 32	18 pts. disseminated CA	13.5-24mg/kg I.V. Push 27-30mg/kg 4 days Inf.	43% 33%	0% 50%

(1) WBC <4000 mmc except Ref. 32 (WBC <3000 mmc)
(2) Mucositis, moderate to severe

With respect to FUra, there are several substances which can theoretically modulate its activity. Chart 2 illustrates three examples of metabolic modulations. (I) Thymidine (dThd) when administered concurrently with FUra, results in reduction of inhibition of DNA synthesis, due to bypass of FdUMP-inhibited dTMP synthesis. At the same time the excess of dTTP and dThd results in reduced formation of the deoxyderivatives FdUDP (by dTTP feedback inhibition of ribonucleotide reductase) and FdUMP (by dThd competition with FdUrd). Therefore the exogenous administration of dThd will direct FUra metabolism towards incorporation into RNA. Under these conditions, the antitumor activity of FUra against some murine solid tumors was increased (17-19). (II) Deoxyinosine (dIno) is a source of deoxyribose-1-phosphate (dR-1-P), which is the deoxyribose 'donor' for the conversion of FUra to FdUrd. Evans et al. demonstrated that administration of dIno to cells in culture increased the FdUMP pool and potentiated the cytotoxicity of FUra (20). (III) With excess folinic acid (dLCF) in culture, the site of action of FUra was shifted from an RNA effect to a more pronounced and prolonged inhibition of thymidylate synthesis and DNA synthesis (20). The concentration of dLCF in the culture medium required to potentiate 3 fold the growth inhibitory effect of FUra in human Hep-2 cells was 20 μM.

In this presentation, a metabolic modulation of FUra by dThd on dLCF that would offer the possibility of increasing the therapeutic efficacy of FUra in experimental systems and in patients with advanced colorectal carcinoma will be discussed.

MATERIALS AND METHODS

FUra was obtained from Hoffman-La Roche, Inc. (Nutley, NJ); Calcium Leucovorin (folinic acid, dLCF) was supplied by Lederle Lab. (Pearl River, NY) and dThd was obtained from the National Cancer Institute (Bethesda, MD). The same substances used as chemical standards were

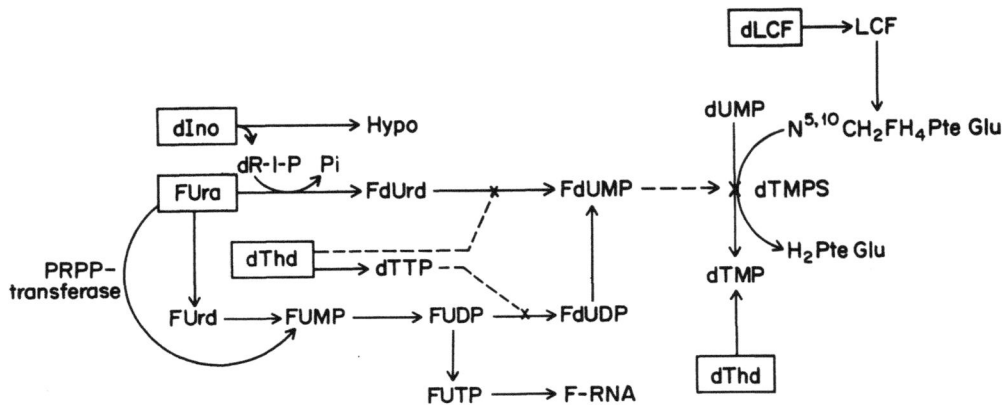

Chart 2. Effects of dThd, dIno and dLCF on the metabolism and sites of action of FUra. broken lines (- - - -) indicate inhibitory effects (see text).

purchased from Sigma Chemical Co. (St. Louis, MO). All other chemical and reagents used were of analytical grade.

Fisher CDF rats (100-150 grams) were obtained from Charles River (Wilmington, MA); Balb/c and C57Bl/6J mice (18-22 grams) were purchased from Jackson Lab. (Bar Harbor, ME). The continuous i.v. infusion of drugs in unrestrained rodents was performed according to the method described by Danhauser and Rustum (22).

For the clinical studies, patient elegibility, protocols and criteria for response have been previously described (23-25). Analysis of the drugs in animal and patient samples has been performed according to published methods (23, 25-27).

RESULTS

(A) FUra + dThd - Preclinical Studies - The effect of different doses of FUra and dThd, alone and in combination, on the survival of normal mice and rats were evaluated following a 72 hr continuous i.v. infusion of FUra (22,26). In rats, dThd alone infused at doses up to 3.3 gm/kg/d x 7 days produced no mortality. The MTD, LD_{50} and LD_{95} for FUra were determined as 50, 107 and 150 mg/kg/d, respectively. The coadministration of 1.7 gm/kg/d dThd and FUra reduced the FUra LD_{50} by approximately 2.5 fold. Furthermore, combinations of 2.5 gm/kg/d dThd with the sublethal dose of FUra (50 mg/kg/d) produced 100% lethality. When analogous experiments were carried out in Balb/c and C57Bl/6J mice, the concurrent administration of dThd reduced the FUra LD_{50} by 40% and 35% respectively (26).

The data in Table 2 summarize the effects of dThd on the MTD of FUra in mice and rats treated by 72 hr continuous i.v. infusion. The MTD was reduced by the concurrent administration of dThd by 28-36%. These results demonstrate that dThd can significantly potentiate the in vivo toxicity of FUra.

Table 2. Effect of coadministration of dThd (1.7 gm/kg/d x 3) on the maximally tolerated dose (MTD) of a 72 hr continuous i.v infusion of FUra to rats and mice.

Animal Strain	MTD (mg/kg/d)	
	FUra	FUra + dThd
Fisher CDF rats	50	33
Balb/c mice	50	33
C57Bl/6J mice	58	42

The results of the antitumor activity of FUra administered alone and in combination with dThd to mice bearing colon tumors No. 26 and 38 and to rats bearing the chemically induced colon carcinoma are summarized in Tables 3 and 4. Drugs were administered by 72 hr continuous i.v. infusion. These results indicate that dThd did not enhance the antitumor activity of FUra at equitoxic doses. If equimolar doses are chosen (Table 4), however, FUra alone produced the same tumor growth inhibition but less tumor free survivors than the combination FUra + dThd : 11% vs. 30%.

Table 3. Antitumor effect of FUra alone and in combination with dThd against mice bearing transplantable colon carcinoma.

Tumor and strain	Treatment FUra (mg/kg/d)	dThd (gm/kg/d)	Tumor growth inhibition (%)	Toxic death (%)
Colon 26 - Balb/c	50	-	90 ± 2[a]	11
	33	1.7	88 ± 5	7
Colon 38 - C5781/6J	58	-	74 ± 4	0
	42	1.7	60 ± 0	0

[a] Mean ± S.D.

Table 4. Antitumor effect of FUra alone and in combination with dThd against rats bearing transplantable colon carcinoma.

Treatment FUra (mg/kg/d)	dThd (gm/mg/d)	Tumor growth inhibition (%)	Tumor free survivors (%)	Toxic death (%)
33	-	97 ± 3[a]	11	0
50	-	98 ± 3	32	14
33	1.7	99 ± 1	30	25

[a] Mean ± S.D.

(B) FUra + dThd - Human Pharmacokinetics and Clinical Studies - The pharmacokinetic of FUra alone and in combination with dThd have been analyzed in patients with advanced colorectal carcinoma (23). Chart 3 illustrates the mean plasma concentrations of FUra in patients receiving 5 days continuous i.v.infusion of FUra alone at the dose of 15 mg/kg/d (I) and in patients receiving 5 days continuous i.v.infusion of FUra 7.5 mg/kg/d + dThd 216 mg/kg/d, preceded by a loading dose of dThd (405 mg/kg by i.v.infusion over 30 minutes) (II). For the infusion of FUra alone the steady-state concentration was reached within one hour and maintained for the duration of the infusion; when dThd was administered in combination with FUra, the latter showed a peak concentration at 24 hr (median value: 15.7 μM) and then decreased to reach its steady-state concentration after 48 hr. The mean values of the steady-state concentration were higher in those patients who received the combination FUra + dThd, despite the lower dose aministered. Moreover, dThd reduced drastically the FUra plasma clearance (data not shown), from 389 ± 154 L/kg/d to 56 ± 36 L/kg/d (two tailed t-test: p<0.001). These results demonstrate that the kinetic behaviours of FUra were altered significantly by dThd.

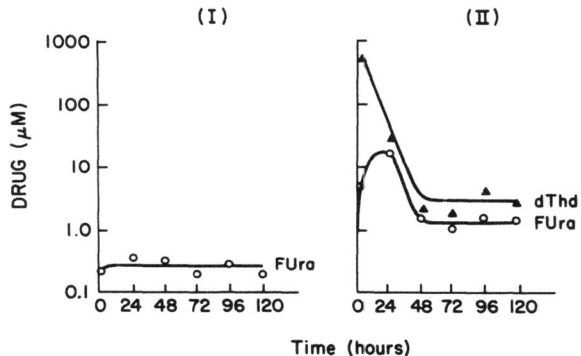

Chart 3. Plasma concentrations of FUra and dThd in patients receiving FUra 15mg/kg/d (I) or FUra 7.5/mg/d + dThd 216mg/kg/d (II) by 5 days continuous i.v. infusion (23).

Spectrum of toxicity and rate of clinical response to the combination FUra + dThd were evaluated in patients with colorectal carcinoma (23,24). The incidence of side effects is outlined in Table 5. With respect to historical controls (Table 1), the combination of FUra + dThd resulted in higher bone marrow toxicity (leukopenia: 64% vs. 12-39%) and reduced G.I.toxicity (mucositis: 19% vs. 33-65%). The G.I. side effects seemed to be partially related to a prior FUra treatment. No complete response were seen among the twenty-two evaluable patients, and the overall response rate was only 5%. These results suggest that the co-administration of FUra and dThd by 5 days continuous i.v.infusion at a FUra dose which is 1/2 of the one employed for the continuous i.v.

infusion of FUra alone resulted in enhanced bone marrow toxicity and failed to improve the antitumor efficacy of FUra. These studies demonstrated that there was a relationship between the plasma FUra concentrations at steady-state and the incidence of bone marrow toxicity. The likelihood of bone marrow toxicity increased with increased plasma steady-state concentration of FUra: 90% (19/21) of treatment courses in which the FUra steady-state level at 120 hr was above 1.5 µM resulted in leukopenia (WBC < 3000 mmc), while this side effect was detected in only 10% (3/29) of the treatment courses with a FUra plateau below 1.5 µM.

Table 5. Incidence of side effects following FUra + dThd by 5 days continuous i.v. infusion to patients with advanced colorectal carcinoma.

Side effect	Overall (%)	Prior FUra (%)	No prior FUra (%)
Leukopenia (< 4000 mmc)	64	69	57
Nausea	50	71	14
Vomiting	31	50	0
Stomatitis	19	14	14

Table 6. Antitumor activity of FUra alone and in combination with dLCF against Balb/c mice bearing colon carcinoma N 26. Drugs were administered by continuous i.v. infusion.

Treatment Drug and dose (mg/kg/d)	Duration of infusion (days)	A.S.T.[a] (days)	M.S.T.[b] (days)	T/C x 100
Saline	2	33.8 ± 12.2	39.0	--
FUra 50	2	33.6 ± 11.0	34.0	87
FUra 100	2	30.4 ± 13.7	29.0	74
dLCF 1500 ↓ FUra 50	1 2	32.0 ± 11.1	29.0	74
dLCF 1500 ↓ FUra 100	1 2	35.3 ± 18.5	45.0	115

[a] = Average Survival Time (mean ± S.D.)
[b] = Median Survival Time

(C) FUra + dLCF - Preclinical studies - The antitumor activity of FUra alone and in combination with dLCF has been investigated in Balb/c mice bearing colon carcinoma No. 26. FUra was given as a continuous i.v. infusion over 2 days. The results are shown in Table 6. No significant antitumor activity was seen following the administration of FUra at doses up to 100 mg/kg/d. Coadministration with high dose dLCF (1.5 gm/kg) did not improve the antitumor efficacy of FUra at the dose of 50 mg/kg/d but did enhance (although slightly) the effectiveness of FUra when used at the dose of 100 mg/kg/d.

(D) FUra + dLCF - Human pharmacokinetics and clinical studies - The combination of FUra and dLCF has been evaluated in patients with advanced colorectal carcinoma receiving the following schedules: FUra (8.1 to 20.3 mg/kg) by i.v. push + dLCF (13.5 mg/kg) p.o. or by 2 hr i.v. infusion (25,28) and FUra (10.8 mg/kg/d) + dLCF (13.5 mg/kg/d) by 5 days continuous i.v. infusion.

Pharmacokinetic studies have been carried out together with the clinical evaluation of toxicity and response. The dose of 13.5 mg/kg dLCF has been chosen in an attempt to achieve plasma levels of active folates > 10 µM, the concentration reported to successfully potentiate the effect of FUra against human tumor cells in culture (20).

Oral administration of dLCF was tested because of the hypothesis of a selective uptake of the active stereoisomer LCF in the gut. In 5 patients receiving the p.o. dLCF the amount of dLCF detected in the plasma was below 1 µM and the amount of its metabolite $5-CH_3-FH_4$ was below 4 µM.

Pharmacokinetic data of FUra administered by i.v. push in the presence of high or low folate plasma levels are shown in Table 7.

Table 7. Effect of folates on the pharmacokinetic parameters of FUra. FUra was administered by i.v.push.

Folate plasma peak (µM)	FUra dose (mg/kg)	T $1/2_\beta$ (min)	Pl. Clearance (mg/kg.min)	Vd (L/kg)
< 5	8.1	9.6 ± 0.8[a]	16.4 ± 5.0	0.22 ± 0.05
134 ± 50	16.2	12.1 ± 2.2	14.0 ± 6.6	0.24 ± 0.13

[a] Mean ± S.D.

These data seem to indicate that in the dose range of 8.1-16.2 mg/kg FUra displayed a dose-dependent kinetic. This may be due to the altered FUra metabolism in the presence of high folate pools in the plasma of those patients who displayed a slower rate of FUra elimination (longer elimination half-life and reduced plasma clearance). The overall response rate in this study was 39%, and among the patients previously resistant to FUra alone the response rate was 50% (Table 8) as reported previously (25). Toxicity included leukopenia and G.I.toxicity, and both appeared to be dose related (Table 9). These results indicate that the coadministration of FUra and dLCF did result in a higher response rate in patients with advanced colorectal cancers.

Table 8. Response of colorectal cancer patients to the combination FUra (i.v.push) + high dose dLCF (2 hr i.v.infusion).

Response	Overall (%)	Prior FUra (%)	No prior FUra (%)
Complete	0	0	0
Partial	39	50	28
Stable Disease	26	17	36
Progression	35	33	36

Table 9. Incidence of side effects following i.v.push FUra + 2 hr i.v.infusion dLCF treatment.

FUra dose (mg/kg)	B.M. Toxicity (WBC <4000 mmc)	G.I. Toxicity (grade 3-4)
8.1	0	0
16.2	17	17
20.3	73	55

Straw et al. have recently demonstrated that after i.v. administration of dLCF the plasma elimination half-life of the active diastereoisomer LCF is 32 min and its plasma clearance is 390 ml/min, while the elimination half-life and plasma clearance of the unnatural d-form are 450 min and 13 ml/min respectively (29). The relatively quick elimination of LCF from the body could therefore lead to the need to administer the dLCF by continuous i.v. infusion. A study of FUra + dLCF by 5 days continuous i.v. infusion has been recently initiated. To date, four patients received one to three cycles of therapy. The preliminary results indicate that the dose of 10.8 mg/kg/d FUra averaged plasma steady-state level of 1.1 µM (0.3-1.5). Steady-state concentrations of LCF and $5\text{-}CH_3\text{-}FH_4$ were of the order of 0.5 - 3 µM and 8 - 20 µM respectively.

DISCUSSION

Large bowel cancers are a major problem in surgical and medical oncology. The survival rate of patients with advanced colorectal carcinoma after surgical excision has not been improved in the past two decades (1). In most cases, at the time of diagnosis cancer has already spread to other organs, therefore a systemic treatment is required to control the minimal residual tumor and the liver and/or lung metastasis.

FUra is, to date, the drug of choice for the treatment of colorectal cancers (1). Unfortunately, response rates greater than 20% are seldom reported. Attempts to improve FUra efficacy by changing the route and schedule of drug administration have been repeatedly carried out (30-32). In terms of clinical response none of these studies have clearly reported any significant advantages over the others even though blood pharmacokinetic modulations of the drugs have been induced. On the other hand, preclinical in vitro (20,21) and in vivo (17) results reported successful modulation of the effects of FUra when pharmacodynamic considerations of drug action have been taken into consideration in the design of in vivo regimens. A clinical counterpart of this is the recent successful use of FUra in combination with high dose dLCF in G.I. cancer patients (25,33).

The results of preclinical and clinical studies involving the metabolic modulation of FUra with dThd and dLCF have been presented herein. The choice of the continuous i.v. infusion as a modality of FUra administration arise from the observation that although the infusion of the drug did not increase the response rate of colorectal cancer patients with respect to the i.v. push treatment, the incidence of bone marrow toxicity and, in fact, of the overall toxicity was reduced (30-32).

Results of the combination of FUra + dThd administered by 72 hr continuous i.v. infusion to tumor bearing animals demonstrate that, on a molar FUra basis, dThd increased the FUra antitumor activity, but also enhanced the overall toxicity (Table 4). When equitoxic doses were chosen, no substantial therapeutic advantage was observed for the FUra + dThd combination (Table 3). These results indicate that in transplanted colon tumors of mice and rats the biological activity of FUra was modulated by dThd, but since both antitumor activity and toxicity were enhanced, the selectivity of this agent was not significantly improved.

A phase I study of the association FUra administered by i.v. push and dThd administered by short term i.v. infusion (10 to 90 min) was carried out by Woodcock et al. (34). Two partial remissions among 18 heavily pretreated colorectal cancer patients (17 of which had progressed under previous FUra-containing regimens) were reported. Despite this unsatisfactory 11% response rate, the fact that two FUra resistant patients did respond to the combination of FUra + dThd was encouraging.

A phase I-II study of FUra + dThd administered by 5 days continuous i.v. infusion was initiated at Roswell Park Memorial Institute. The effects of dThd on the pharmacokinetic properties of FUra were evaluated. Administration of dThd produced prolongation of the FUra plasma half-life and reduced the plasma clearance of the drug by nearly 7 folds (23). The response rate observed in this study was even lower with respect to the value reported by Woodcock: one partial response in 22 evaluable patients was seen (24). Furthermore, the incidence of side effects was high, with leukopenia experienced by 64% of patients (Table 5). In addition, a shift in profile of toxicity from G.I. toxicity with FUra alone to bone marrow toxicity with FUra + dThd was seen (Table 1). This was especially the case in those patients whose FUra plasma steady-state levels were greater than 1.5 μM. The results of this study indicated that the increase of FUTP-mediated RNA effects by the concurrent administration of dThd failed to improve the clinical response rate of advanced colorectal cancer patients. If one is to assume that the lack of clinical effectiveness of this association could be related to a decrease of FdUMP-mediated DNA effects, the coadministration of reduced folate and FUra might increase the clinical response to the antimetabolite.

Laskin et al. have demonstrated that human tumor cell lines in general are less sensitive to FUra than mouse tumor cell lines (35). Differences in FUra sensitivity were correlated with differences in drug metabolism. For example, human carcinoma Hep-2 cells are far less sensitive to FUra than mouse sarcoma S-180 (35). Furthermore, while dTMP-S inhibition is the growth limiting target of FUra in S-180, incorporation of FUra into RNA is the main site of action in Hep-2 cells (36). The results of these studies suggests that determinants of response for one cell type may not be the same for other cell types. Moreover, dependent on the intracellular biochemical make of each cell type, human cells in culture appears to be significantly different from mouse cells.

For example cells such as Hep-2 cells have significantly lower levels of reduced folate than mouse cell such as S-180 (21). Thus, the poorly FUra sensitive Hep-2 cells can be made highly sensitive by the coadministration of FUra and excess folate. Hence, sensitivity of human cells became comparable to that of S-180 cells when the intracellular folate pools were modified to equal levels (20,21). These in vitro results demonstrated that intracellular folate pool concentrations appear to play a major role in determining sensitivity to FUra.

In vivo preclinical results of FUra + dLCF against mouse bearing colon carcinoma No. 26 failed to show significant enhancement of the antitumor efficacy of FUra. These results indicated, however, that dLCF did not potentiate overall host toxicity to FUra. Based on the strong rational developed in in vitro studies and on the lack of in vivo increase of host toxicity, a clinical Phase I-II protocol was designed at Roswell Park Memorial Institute. The results of this phase I-II trial have been reported (25). Similar findings have been published by Machover et al. (33). In brief, an overall response rate of 39% was observed (Table 8) and 6/12 of patients previously resistant to other chemoterapeutic regimens containing FUra responded to the FUra + high dose dLCF treatment. At the dose of 13.5 mg/kg dLCF, a peak plasma level of folate above 100 μM was achieved (Table 7). The plasma peak of the biological active form (L-CF) was in the range of 9 to 30 μM. Due to its short half-life, plasma concentrations of LCF may not be maintained in the systemic circulation for a long enough time to modulate optimally the FUra effects. For this reason a phase I-II trial of high dose dLCF administered by 5 days continuous i.v. infusion in combination with FUra had been initiated: the dose of 13.5 mg/kg/d dLCF provided long-term plasma concentrations of active folates in the range of 10 to 20 μM.

In conclusion:
(1) The combination of FUra and dThd failed to improve the selectivity of FUra in experimental models and in human colorectal carcinoma despite strong evidence of the in vitro biochemical rationale for the modulatory activity of this combination.
(2) The combination of FUra and high doses dLCF seems promising in patients with advanced colorectal cancers.

(3) The continuous i.v. infusion of FUra, when resulting in plasma steady-state concentrations in the order of 1.0 - 1.5 µM, is as effective as the i.v. push administration and is relatively safer, with a low incidence of bone marrow toxicity.

(4) Increased antitumor efficacy of an antimetabolite like FUra can be induced also in humans through the application of the concept of metabolic modulation.

REFERENCES

1. Carter S.K. Large bowel cancer - the current status of treatment. J. Natl. Cancer Inst. 56:3-10, 1976.
2. Levin P., Mittelman A., Douglass H., Engstrom P., Klaassen D. Survival and response rate to chemotherapy for advanced colorectal adenocarcinoma: an Eastern Cooperative Oncology Group report. Cancer (Phila.) 46:1536-1543, 1980.
3. Cancer - Principles and practice of oncology. De Vita V.T.Jr., Hellman S., Rosenberg A. editors, J.B. Lippincott Co., Philadelphia 1982.
4. Mandel H.G. The incorporation of 5-fluorouracil in RNA and its molecular consequences. Prog. Mol. Subcell. Biol. 1:82-135, 1969.
5. Glazer R.I., Peale A.L. The effect of 5-fluorouracil on the synthesis of nuclear RNA in L1210 cells in vitro. Mol. Pharmacol. 16:270-277, 1979.
6. Kufe S.W., Major P.P. 5-Fluorouracil incorporation into human breast carcinoma DNA correlates with cytotoxicity. J. Biol. Chem. 256:9802-9805, 1981.
7. Glazer R.I., Lloyd L.S. Association of cell lethality with incorporation of 5-fluorouracil and 5-fluorouridine into nuclear RNA in human colon carcinoma cells in culture. Mol. Pharmacol. 21:468-473, 1982.
8. Hartman K.U., Heidelberg C. Studies on fluorinated pyrimidines. XIII. Inhibition of thymidylate synthetase. J. Biol. Chem. 236:3006-3013, 1961.
9. Santi D.V., McHenry C.S., Sommer H. Mechanism of interaction of thymidylate synthetase with 5-fluorodeoxyuridilate. Biochemistry 13:471-480, 1974.
10. Danenberg P.V., Locksin A. Fluorinated pyrimidines as tight-binding inhibitors of thymidylate synthetase. Pharmacol. Ther. 13:69-90, 1981.
11. Houghton J.A., Maroda S.J., Phillips J.D., Houghton P.J. Biochemical determinants of responsiveness to 5-fluorouracil and its derivatives in xenografts of human colorectal adenocarcinoma in mice. Cancer Res. 41:144-149, 1981.
12. Major P.P., Egan E., Herrick D., Kufe D.F. 5-Fluorouracil incorporation in DNA of human breast carcinoma cells. Cancer Res. 42:3005-3009, 1982.
13. Schuetz J.D., Wallace H.J., Diasio R.B. 5-Fluorouracil incorporation into DNA of CF-1 mouse bone marrow cells as a possible mechanism of toxicity. Cancer Res. 44:1358-1363, 1984.
14. Sawyer R.C., Stolfi R.L., Martin D.S., Spiegelman S. Incorporation of 5-fluorouracil into murine bone marrow DNA in vivo. Cancer Res. 44:1847-1851, 1984.
15. Lonn U., Lonn S. Interaction between 5-fluorouracil and DNA of human colon adenocarcinoma. Cancer Res. 44:3414-3418, 1984.

16. Bloch A. Metabolic Conditioning and Metabolic Actuation: experimental approaches to cancer chemotherapy involving combinations of metabolites and antimetabolites. Cancer Chemother. Rep. 58:471-477, 1974.
17. Martin D.S., Stolfi R.L., Spiegelman S. Striking augmentation of the in vivo anticancer activity of 5-fluorouracil by combination with pyrimidine nucleosides: an RNA effect. Proc. Am. Assoc. Cancer Res. 19:221, 1978.
18. Nayak R., Martin D.S., Stolfi R.L., Furth J., Spiegelman S. Pyrimidine nucleosides enhance the anticancer activity of 5-fluorouracil and augment its incorporation into nuclear RNA. Proc. Am. Assoc. Cancer Res. 19:1963, 1978.
19. Martin D.S., Stolfi R.L., Sawyer R.C., Nayak R., Spiegelman S., Young C.W., Woodcock T. An overview of thymidine. Cancer (Phila.) 45:1117-1128, 1980.
20. Evans R.M., Laskin J.D., Hakala M.T. Effect of excess folates and deoxyinosine on the activity and site of action of 5-fluorouracil. Cancer Res. 41:3288-3295, 1981.
21. Yin M-B., Zakrzewski S.F., Hakala M.T. Relationship of cellular folate cofactor pools to activity of 5-fluorouracil. Mol. Pharmacol. 23:190-197, 1983.
22. Danhauser L.L., Rustum Y.M. A method for continuous drug infusion in unrestrained rats: its application in evaluating the toxicity of 5-fluorouracil/thymidine combination. J. Lab. Clin. Med. 93:1047-1053, 1979.
23. Au J.L-S., Rustum Y.M., Ledesma E.J., Mittelman A., Creaven P.J. Clinical pharmacological studies of concurrent infusion of 5-fluorouracil and thymidine in treatment of colorectal carcinomas. Cancer Res. 42:2930-2937, 1982.
24. Sternberg A., Petrelli N.J., Au J., Rustum Y., Mittelman A., Creaven P.A combination of 5-fluorouracil and thymidine in advanced colorectal carcinoma. Cancer Chemother. Pharmacol. 13:218-222, 1984.
25. Madajewicz S., Petrelli N., Rustum Y., Campbell J. Herrera L., Mittelman A., Perry A., Creaven P.J. Phase I-II trial of high-dose calcium leucovorin and 5-fluorouracil in advanced colorectal cancer. Cancer Res. 44:4667-4669, 1984.
26. Danhauser L.L., Rustum Y.M. Chemotherapeutic efficacy of 5-fluorouracil with concurrent thymidine infusion against transplantable colon tumor in rodents. Cancer Drug Delivery 1(4): 269-282, 1984.
27. Rustum Y.M. High pressure liquid chromatography. I. Quantitative separation of purine and pyrimidine nucleosides and bases. Anal. Biochem. 90:289-299, 1978.
28. Trave F., Rustum Y.M., Mazzoni A., Petrelli N., Madajewicz S., Mittelman A., Creaven P. Possible effect of high dose 5-formyltetrahydrofolic acid on the pharmacokinetics of 5-fluorouracil in patients with colorectal carcinoma. Proc. Am. Soc. Clin. Oncol. 4:30, 1985.
29. Straw J.A., Szapary D., Wynn W.T. Pharmacokinetic of the diastereoisomers of leucovorin after intravenous and oral administration to normal subjects. Cancer Res. 44:3114-3119, 1984.
30. Moertel C.G., Schut A.J., Reitemeier R.J., Hahn R.G. A comparison of 5-fluorouracil administered by slow infusion and rapid injection. Cancer Res. 32:2717-2719, 1972.
31. Seifert P., Baker L.H., Reed M.L., Vaitkevicius V.K. Comparison of continuously infused 5-fluorouracil with bolus injection in treatment of patients with colorectal adenocarcinoma. Cancer 36:123-128, 1975.
32. Fraile R.J., Baker L.H., Buroker T.R., Horwitz J., Vaitkevicius V.K. Pharmacokinetics of 5-fluorouracil administered orally, by rapid intravenous and by slow infusion. Cancer Res. 42:2223-2228, 1980.

33. Machover D., Schwarzenberg L., Goldschmidt E., Tourani J.M., Michalski B.Hayat M., Dorval T., Misset J.L., Jasmin C., Maral R., Mathe' G. Treatment of advanced colorectal and gastric adenocarcinomas with 5-FU combined with high dose folinic acid: a pilot study. Cancer Treat. Rep. 66:1803-1807, 1982.
34. Woodcock T.M., Martin D.S., Damin L.A.M., Kemeny N.E., Yuong C.W. Combination clinical trials with thymidine and fluorouracil: a phase I and clinical pharmacologic evaluation. Cancer 45:1135-1143, 1980.
35. Laskin J.D., Evans R.M., Slocum H.K., Burke D., Hakala M.T. Basis for natural variation in sensitivity to 5-fluorouracil in mouse and human cells in culture. Cancer Res. 39:383-390, 1979.
36. Evans R.M., Laskin J.D., Hakala M.D. Assessment of growth limiting events caused by 5-fluorouracil in mouse cells and in human cells. Cancer Res. 40:4113-4122, 1980.

ACKNOWLEDGEMENTS

This work has been supported in part by CA 21071 and CA 18420 from the Department of Health and Human Services, and by the US-Italy agreement on Cancer Research (F.T.).

EPIPODOPHYLLOTOXIN AND CISPLATIN

ON CONTINUOUS INFUSION SCHEDULES

>   Jacob J. Lokich
>
>   Chief, Clinical Oncology
>   New England Deaconess Hospital
>   Harvard Medical School
>   Boston, Massachusetts

The Epipodophyllotoxins, VP16-213 and VM26, and the heavy metal cytotoxic agents, Cisplatin and its analogs, Spirogermanium and Gallium, represent two classes of agents which in clinical trials are traditionally delivered on an intermittent bolus schedule. Extensive clinical reviews of the clinical trials employing these agents have not emphasized the continuous infusion schedule (1-3). In fact, the thrust has been directed toward maximizing the dose of delivery on an intermittent bolus schedule to increase therapeutic effects (4,5). Such has been the traditional approach to cancer chemotherapy in general based upon the concept of the dose-response relationship developed in experimental tumor systems and upon practical issues involving patient convenience and outpatient delivery. Such precepts have been the basic tenets for the day one and eight schedule for such programs as MOPP chemotherapy for Hodgkin's disease and the CMF program for breast cancer.

The continuous infusion schedule emphasized the concept of concentration over time to maximize tumor cell killing and is based upon the pharmacokinetics of chemotherapeutic agents; cytokinetic considerations of cell growth; and in vivo and in vitro experimental tumor systems demonstrating schedule dependency for most chemotherapeutic agents.

Infusion chemotherapy may be separated into three categories based upon the duration of the infusion, as well as the route of delivery. Infusions administered for less than twenty-four hours, generally for one to six hours, should be comparable pharmacokinetically to bolus delivery and are generally applied for the purpose of avoiding local vascular damage during the infusion or peak effects such as hypotension or CNS disturbances. The three types of infusion are short-term systemic infusion for ninety-six to one hundred and twenty hours; protracted systemic infusion for fourteen to twenty-eight days; and regional infusion, most commonly to the liver, administered for fourteen to twenty-one days consecutively. The rationale for each of the three categories is distinctive. For the short-term infusion the goal is to deliver a maximum tolerable dose over the defined interval with courses repeated at 4-5 weeks. Ninety-six hour infusions are distinguished from 120 hour infusions for practical logistical reasons, permitting termination of the shorter duration on a four day schedule (Monday through Friday) to coincide with radiation therapy. Protracted systemic infusion is based upon maximizing the duration of exposure with a theoretic goal of encompassing a tumor doubling time, therapy maximizing exposure of the sensitive tumor cell population composed of the cells entering cycle. Regional infusion permits delivery of a maximal dose locally with minimal host toxicity as a consequence of drug extraction within the region.

The major impetus to the development and application of infusion schedules to cancer chemotherapy in general has been the technologic evolution of safe and reliable vascular access and the development of portable and accurate delivery pumps. It is beyond the intended scope of this discussion to review these technologic advances, but the central role of these developments cannot be overemphasized. Details of the surgical placement and complications of the vascular access devices (6,7) and the portable infusion devices (8) have been published.

EPIPODOPHYLLOTOXINS

The clinically available epipodophyllotoxins include Etoposide (VP16-213) and Teniposide (VM26) which are analogs of the podophyllotoxin and are natural products derived from the mandrake root. Etoposide has demonstrated activity in small cell carcinoma of the lung, non-Hodgkin's lymphoma, germ cell tumors and acute leukemia (9). Although generally employed on a standard intermittent bolus schedule, preclinical studies in L1210 leukemia has demonstrated marked schedule dependency for this agent (10), supporting a rationale for continuous infusion. In addition, pharmacokinetic studies have demonstrated a relatively short plasma half-life of approximately one hour (T one-half alpha) and five hours (T one-half beta) (11). Clinical trials employing a variety of schedules with bolus delivery in small cell carcinoma indicate a higher response rate with a consecutive or every other day regimen compared to a weekly regimen (12). Thus, continuous infusion as a schedule for this agent is supported at both the preclinical and the clinical level. Teniposide demonstrates pharmacokinetic parameters similar to Etoposide (13) which would support a continuous infusion schedule for this agent as well, but clinical trials have not been carried out as extensively as with VP16-312.

There are, in fact, only three trials in which continuous infusion of Etoposide has been studied (Table 1). This may be related in part to the fact that Etoposide is a relatively insoluble agent and requires large volumes as a diluent in order to maintain the drug in solution, thus making outpatient infusion systems logistically problematic. This important pharmaceutical consideration also accounts for the fact that only short-term continuous infusion trials for five days have been studied.

TABLE 1

VP16 - 213   CONTINUOUS INFUSION TRIALS

Study	Design	No. Patients	Dose Schedule
Aisner et al. CCP 1982	Phase 1 Breast	17	75-150 $mg/M^2/d$ x 5d *
Schell et al. CCP 1982	Phase 111 Breast	77	50-70 $mg/M^2/d$ x 5d
Lokich et al. CTR 1979	Phase 1	24	20-80 $mg/M^2/d$ **

Recommended daily dose rate
* 125 $mg/M^2/d$
** 60 $mg/M^2/d$

Two of the continuous infusion trials represent Phase I studies designed to establish the optimal dose of delivery (14-15). In the study by Aisner et al 17 patients were studied at dose rates of 75 mg/M^2/d; 100 mg/M^2/d; 125 mg/M^2/d and 150 mg/M^2/d. Dose limiting toxicity was manifest as bone marrow suppression and stomatitis was also observed. In addition, adverse cardiac effects were observed in 3 patients. The authors concluded that the optimal dose rate of delivery was 125 mg/M^2/d for five days. Two responses were observed in this Phase I trial in a patient with renal cell carcinoma and in a patient with seminoma.

In the Phase I trial by Lokich et al 24 patients were studied in a five day continuous infusion program employing daily dose rates of 20 mg/M^2 to 80 mg/M^2 per day. Dose limiting toxicity was bone marrow suppression with stomatitis also demonstrated at the higher dose rates. The recommended dose rate of delivery for this trial was 60 mg/M^2/d for five days. The difference in recommended dose rates of delivery for these two Phase I trials is substantial with the trial by Aisner et al recommending twice the daily dose rate as the trial by Lokich et al. This difference may be related to differences in the patient mix or other factors not readily apparent.

The only Phase III trial reported comparing the infusion and bolus schedule for Etoposide was carried out in patients with metastatic breast cancer (16). The continuous infusion schedule involved five days of drug administration at a dose of 50-70 mg/M^2/d, precisely similar to the recommendation of Lokich et al. Tumor responses were observed in 14 per cent of patients receiving bolus VP16-213 and in 13 per cent of patients receiving continuous infusion schedule for this drug. One complete response was observed in a patient receiving the continuous infusion schedule. Although the observed response rate is small, it should be noted that all patients had received extensive prior therapy with more than one combination chemotherapy regimen; thus, although there was no apparent therapeutic advantage for the continuous infusion schedule for VP16-213 in this study, one could consider using this agent in previously untreated patients or minimally treated patients, or even in combination with other established active agents for metastatic breast cancer.

In summary, the continuous infusion of VP16-213 has a strong rationale based upon pharmacokinetic considerations of the drug and schedule dependency in experimental tumor systems. Pharmaceutical considerations relative to drug stability, however, limit the easy application of this schedule for general use. Therapeutic effects are observed, permitting at least the conclusion that the infusion schedule does not compromise drug activity, but the true quantitative level of activity cannot be determined on the basis of the limited available infusion trials. Tumor response have been observed in breast cancer and in two of two patients with non-Hodgkin's lymphoma in the trial by Lokich et al. It may be worthwhile to consider examining continuous infusion VP16-213 in the context of a larger Phase II trial to determine if the spectrum of tumors for which this drug may be considered active could be expanded. In addition, the use of the infusion schedule for this agent in conjunction with multidrug programs may permit delivery of more effective doses of the other component agents by limiting the compounding toxicity of VP16-213.

The continuous infusion schedule did not substantially alter the pattern of toxicity with the exception of the observation that stomatitis was often observed on the infusion schedule and bone marrow suppression was the major dose limiting toxicity. In a recently reported multidrug regimen incorporating cyclophosphamide and adriamycin along with continuous infusion VP16-213 for small cell carcinoma of the lung, substantial bone marrow suppression was observed resulting in two drug-related deaths

(17), suggesting that the continuous infusion schedule offered no advantage with regard to toxicity. In that same study the therapeutic effects were comparable to those reported for the bolus schedule.

CISPLATIN

Cis-diamminedichloroplatinum or Cisplatin is the first heavy metal compound to enter clinical trials and has demonstrated substantial activity, particularly in testicular and ovarian cancers, but also in epidermoid tumors of the aerodigestive tract, gastric cancer, and lung cancer. Early Phase I trials had identified extraordinary emetogenic effects of the drug and substantial renal toxicity, as well as ototoxicity and peripheral neuropathy. The mechanism of Cisplatin-associated tumor cell killing is via inhibition of DNA synthesis by crosslink binding.

The traditional delivery schedule for Cisplatin is at a dose of 50-120 mg/M^2 as a bolus repeated at 3-4 week intervals. In some studies in an effort to decrease renal toxicity and particularly the marked gastro-intestinal effects, the dose has been distributed over a five day interval generally at a dose of 20 mg/M^2/d. In combination chemotherapy studies the usual dose administered as an outpatient is 50 mg/M^2. In the high dose Cisplatin studies in association with hypertonic saline, a cumulative dose of 200 mg/M^2 was achieved distributing the bolus over five days at a dose of 40 mg/M^2/d (4).

A continuous infusion schedule for Cisplatin is supported on the basis of: (1) preclinical data in experimental in vitro system in which Cisplatin demonstrated an increased antitumor effect with a low dose constant exposure (18); (2) drug pharmacokinetics demonstrating a triphasic decay with terminal half-life that extends to 24 hours (19); and (3) the inordinate gastrointestinal and renal toxicity which could potentially be obliviated by an infusion schedule.

Like VP16-213 there are pharmaceutical issues to be considered in the delivery of Cisplatin as a continuous infusion. The agent is stable upon reconstitution for 5-6 days without degradation, but it is essential that a high chloride concentration be maintained in order to prevent aquation with release of the chloride irons and consequent precipitation of the agent.

The clinical trials employing an infusion schedule for platinum reported to date have been relatively limited (Table 2). Of the four trials, one represents a Phase I trial of the continuous infusion schedule delivered for five days (20); two are Phase II trials (21,22) and the last is another Phase I trial of a protracted infusion for 28 days (23). The study by Gasparini is included in this review in spite of the fact it represents the delivery of the drug only for a 24 hour period (21). In this study the infusion schedule was applied in the Phase II design for treatment of advanced osteogenic sarcoma. The study identified important activity for Cisplatin in osteogenic sarcoma with a response rate of 19% (7/37). However, toxicity did not appear to be modified by the 24 hour infusion in that renal effects were observed in 16% of patients, and clinically significant hearing loss was identified. The emetogenic effects of Cisplatin were not commented upon.

In the Phase I study by Lokich et al employing the five day infusion, 30 patients were entered at dose rates between 20 and 40 mg/M^2/d (20). The dose limiting toxicity was nausea and vomiting, as well as marrow suppression at dose rates greater than 30 mg/M^2/d. Renal toxicity was observed in approximately 10% of patients, but was generally transient and it was suggested that renal effects were substantially reduced by the infusion schedule since there were no concomitant hydration programs de-

TABLE 2

CDDP: CONTINUOUS INFUSION - 120h

Study	No.	Courses	Dose	Schedule
Gasparini et al. CTR 69:211,1985	37	-	100 mg/M^2/d x 1	q 3 wks
Lokich et al. CTR 64:905,1980	30**	41	20-40 mg/M^2/d x 5	q 4-6 wks
Salem et al. Cancer 53:837,1984	96 (20)*	280	20 mg/M^2/d x 5	q 4-6 wks
Lokich et al. CDDI:247,1984	14**	15	5-10 mg/M^2/d x 28	

* Number receiving CDDP alone
** Phase I study

livered with the infusion, and two of the three patients developing renal failure had limited renal reserve prior to initiation of the infusion. Tumor responses were observed in one patient with nonsmall cell carcinoma of the lung and one patient with metastatic germ cell tumor. Although the proposed recommended dose for that study was 30 mg/M^2/d for five days, subsequent experience at a dose of 20 mg/M^2/d has been deemed to be more tolerable.

The Phase II study by Salem et al delivered Cisplatin as a five day continuous infusion at the dose of 20 mg/M^2/d (22). Although the clinical trial is large in terms of numbers of patient entries (96), only 20 patients received Cisplatin as a single agent. The author suggested that there was a major modification of nephrotoxicity (15%), although in this study hydration was an essential part of the regimen. Therapeutic activity was identified, particularly for thymoma, although the specific quantitative level of activity across tumors could not be reasonably assessed. There were also two responses in head and neck cancer.

A protracted infusion schedule is practical for Cisplatin because large volumes are not necessary to maintain the drug in solution as is the case with VP16 and the drug is stable for 4-6 days. In a Phase I study by Lokich et al, dose rates of 5-10 mg/M^2/d for 28 days were evaluated (23). The optimal dose rate appeared to be 5 mg/M^2/d above which intractable nausea occurred. On the protracted infusion schedule 3 of 5 patients developed renal failure predominantly related to compromised renal function prior to entry on the trial. A tumor response was observed in one patient with metastatic carcinoma of thyroid.

In summary, the limited studies of continuous infusion Cisplatin delivered systemically have at least suggested that the therapeutic activity is not compromised as a consequence of the infusion schedule, although it is difficult to make a quantitative assessment or a comparative assessment to the therapeutic efficacy of bolus delivery schedules in the absence of direct Phase III comparative trials. In addition, there have been insufficient patients in any one tumor category to determine in a quantitative sense the response rate for that tumor, and in the largest Phase II study to date the infusion Cisplatin was delivered in conjunction with other agents. Lokich et al have evaluated Cisplatin in a five day infusion delivered concomitantly with oral cyclophosphamide in a Phase II trial in ovarian cancer, and a comparable response rate to that reported for the bolus schedule has been observed (24). In terms of expanding the spectrum of tumors which may be affected by Cisplatin in the context of an infusion schedule, a single unreported trial of Cisplatin by protracted infusion in patients with advanced measurable colorectal cancer failed to identify any activity for this agent on the infusion schedule, similar to the experience for bolus Cisplatin in advanced colorectal cancer (21).

In terms of toxicity, infusional Cisplatin decreases the frequency of renal failure in spite of the absence of vigorous hydration. This observation may in part be related to the fact that the infusion schedule decreases the emetogenic effect of Cisplatin and thereby may obviate intravascular dehydration in the acute setting, preventing renal failure. Neuropathy and ototoxicity, however, may be increased in frequency in patients receiving the infusion schedule. In the Phase I studies of both short-term or protracted infusion, there were insufficient patients receiving multiple courses of therapy to assess the neuropathic effects of Cisplatin which are generally attributed to cumulative drug. However, in the Phase II study in advanced ovarian cancer the frequency of ototoxicity and profound neuropathy approached 30 per cent or more, related most likely to the cumulative drug over time.

Regional delivery of Cisplatin has been reported, although generally as a bolus and not as a continuous infusion (Table 3). The two major organ sites for arterial delivery have been directed at brain tumors and liver metastases. Two studies in intracarotid delivery for brain tumors are provocative (26,27). In the study by Stewart et al, the dose of delivery was 100 mg/M^2 delivered over one hour and responses were observed in 6 of 11 patients. In the study by Lelane et al reported more recently a dose of 100 mg/M^2 was delivered over one hour and an 80 per cent or greater response rate was observed in both primary and metastatic patients to the brain. Toxicity included deafness, visual disturbance and focal and generalized CNS effects with seizure activity.

In the two studies in which intra-arterial Cisplatin was delivered to the liver, all patients had metastatic breast cancer and in one of the studies the drug was, in fact, infused over a 2-5 day period. In the study by Salem et al two of three patients demonstrated an antitumor effect. In the study by Fleishman, 4 of 21 patients demonstrated response. Although the drug was only infused over a 3-4 hour period, but doses up to 150 mg/M^2 were administered.

TABLE 3

CISPLATIN ARTERIAL DELIVERY

Stewart et al. CR 1982	Brain	1° and 2°	60-100 mg/M^2 over 1h	6/11
Lelane CDD 1983	Brain	1° and 2°	100 mg/M^2 over 1h	8/10 * 11/13 **
Salem et al. PAACR 1981	Liver	Breast	100 mg/M^2 over 2-5d	2/3
Fleishman et al. PAACR 1981	Liver	Breast	80-150 mg/M^2 over 4h	4/21

* Primary
** Metastatic

In summary, the regional delivery of Cisplatin has demonstrated some exceptional therapeutic effects, particularly in brain tumors, although toxicity has been substantial, and in fact the infusion schedule was not employed. Regional delivery to the liver using an infusion schedule does not demonstrate a selective advantage based upon pharmacologic studies in which peripheral venous plasma levels are quantitatively comparable to that delivered on standard bolus schedules. Therapeutic efficacy is demonstrable, but with the exception of VP16-213 in advanced metastatic breast cancer prospective comparative trials have not been performed.

VP16-213 has practical limitations with regard to the infusion schedule relative to the insolubility of the agent in small volumes. Nonetheless, the pharmacokinetic characteristics of the agent, the experimental drug trials demonstrating schedule dependency, and the clinical trials to date have all suggested that an infusion schedule or frequent administration will optimize the therapeutic effects, as well as alter toxicity. For Cisplatin the rationale for a continuous infusion schedule is weaker in that schedule dependency is difficult to demonstrate in experimental tumor systems and pharmacologic studies have demonstrated a terminal half-life for this agent of 24 hours. Therefore, based upon such information a frequent bolus administration would be as reasonable as a continuous infusion schedule, although the latter may in some instances be more logistically practical. The substantial amelioration of the gastrointestinal effects of Cisplatin on the infusion schedule must be balanced against the possible increase in the neuropathic effects and hematologic effects with cumulative doses of the drug delivered on this schedule.

REFERENCES

1. P.J. Loehrer, L.H. Einhorn, Cisplatin, Ann Int Med 100:704 (1984).
2. P.J. O'Dwyer, B. Leyland-Jones, M.T. Alonso, Etoposide (VP-16-213): Current Status of an Active Anticancer Drug, NEJM.
3. VP-16: Recent Advances and Future Prospects, Sem Onc XII, No. 1, Suppl 2 (1985).
4. R.F. Ozols, B.J. Corden, J. Jacob, M.N. Wesley, High-Dose Cisplatin in Hypertonic Saline, Ann Int Med 100:19 (1984).
5. S.N. Wolff, D.H. JOhnson, K.R. Hande, High-Dose Etoposide as Single Agent Chemotherapy for Small Cell Carcinoma of the Lung, Can Treat Rep 67:957 (1983).
6. J.J. Lokich, A. Bothe, N. Fine, J. Perri, The Delivery of Cancer Chemotherapy by Constant Venous Infusion: Ambulatory Management of Venous Access and Portable Pump, Cancer 50:2731 (1982).
7. A. Bothe, W. Piccione, J.J. Ambrosino, P.N. Benotti, J.J. Lokich, Implantable Central Venous Access System, Amer J Surg, 147:565 (1984).
8. J.J. Lokich, W. Ensminger, Ambulatory Pump Infusion Devices for Hepatic Artery Infusion, Sem Onc 10(2):183 (1983).
9. H. Schmoll, Review of Etoposide Single-Agent Activity, Can Treat Rep 9(A):21 (1982).
10. W. Achterrath, N. Niederle, R. Raettig, P. Hilgrad, Etoposide - Chemistry, Preclinical and Clinical Pharmacology, Can Treat Rep 9(A):3 (1982).
11. P.J. Loehrer, L.H. Einhorn, Cisplatin, Ann Int Med 100:704 (1984).
12. H. Schmoll, Review of Etoposide Single-Agent Activity. Can Treat Rep 9(A):21 (1982).
13. M. D'Incalci, C. Rossi, C. Sessa, R. Urso, Pharmacokinetics of Teniposide in Patients with Ovarian Cancer, Can Treat Rep 69 (1985).
14. J. Aisner, D.A. VanEcho, M. Whitacre, P.H. Wiernik, A Phase I Trial of Continuous Infusion VP16-213 (Etoposide)*, Can Chemo Pharmacol 7:157 (1982).

15. J.J. Lokich, J. Corkery, Phase I Study of VP-16-213 Administered as a Continuous Five-Day Infusion, Can Treat Rep 65(6-10):887 (1981).
16. F.C. Schell, H.Y. Yap, G.N. Hortobagyi, B. Issell, L. Esparza, Phase II Study of VP16-213 (Etoposide) in Refractory Metastatic Breast Carcinoma*, Can Chemo Pharmacol 7:223 (1982).
17. H.W. Matelski, J.J. Lokich, M.S. Huberman, T.E. Zipoli, Adriamycin, Cyclophosphamide, and Etoposide (VP-16-213) in Extensive-Stage Small Cell Lung Cancer, Am J Clin Oncol 7:729 (1984).
18. B. Drewinko, B.W. Brown, J.A. Gottlieb, The Effect of Cis-Diamminedichloroplatiunum (II) on Cultured Human Lymphoma Cells and its Therapeutic Implications, Can Res 33:3091 (1973).
19. J.J. Roberts, Cisplatin, In: H.M. Pinedo, ed. Cancer Chemotherapy Excerpta Medica, Amsterdam, 1982.
20. J.J. Lokich, Phase I Study of Cis-Diamminedichloroplatinum (II) Administered as a Constant 5-Day Infusion, Can Treat Rep 64:905 (1980).
21. M. Gasparini, J. Rouesse, A. vanOosterom, T. Wagener, Phase II Study of Cisplatin in Advanced Osteogenic Sarcoma, Can Treat Rep 69 (1985).
22. P. Salem, M. Khalyl, K. Jabboury, L. Hashimi, Cis-Diamminedichloroplatinum (II) by 5-Day Continuous Infusion, Cancer 53:837 (1984).
23. J.J. Lokich, T.E. Zipoli, Phase-I Study of Protracted Infusion of Cisplatin, Can Drug Deliv 1 (1984).
24. J.J. Lokich, T.E. Zipoli, Cisplatin Five-Day Infusion Combined with Bolus Oral or Intravenous Cyclophosphamide in Ovarian Cancer, Unpublished observations.
25. J.J. Lokich, Protracted 28-Day Infusion Cisplatin as Second Line Therapy for Advanced Colorectal Cancer, Unpublished observations.
26. D.J. Stewart, S. Wallace, L. Feun, M. Leavens, A Phase I Study of Intracarotid Artery Infusion of Cis-Diamminedichloroplatinum (II) in Patients with Recurrent Malignant Intracerebral Tumors, Can Res 42:2059 (1982).
27. D.E. Lehane, R.N. Bryan, B. Horowitz, L. DeSantos, Intraarterial Cis-Platinum Chemotherapy for Patients with Primary and Metastatic Brain Tumors, Can Drug Deliv 1 (1983).
28. P.A. Salem, M. Khalil, G. Rizk, K. Jabboury, Intra-Hepatic Artery Infusional Chemotherapy with Cis-Platinum in the Treatment of Metastatic Liver from Breast Cancer, AACR Abstracts (1981).
29. G.B. Fleishman, H.Y. Yap, A. DiStefano, G.R. Blumenschein, Intrahepatic Arterial Cis-Diammine-Dichloro Platinum (II) (IHACP) and Vinblastine (IHAV) for Refractory Metastatic Breast Carcinoma (MBC) Confined to the Liver, AACR Abstracts (1981).
30. D.J. Stewart, R.S. Benjamin, S. Zimmerman, R.M. Caprioli, Clinical Pharmacology of Intraarterial cis-Diamminedichloroplatinum (II), Can Res 43:917 (1983).
31. R.P. Warrell, C.J. Coonley, D.J. Straus, C.W. Young, Treatment of Patients with Advanced Malignant Lymphoma Using Gallium Nitrate Administered as a Seven-Day Continuous Infusion, Cancer 51:1982 (1983).

BIODEGRADABLE STARCH MICROSPHERES (SPHEREX), A CLINICALLY USEFUL MEDICAL DEVICE FOR COMBINED INTRA-ARTERIAL CHEMOTHERAPEUTIC TREATMENT OF PRIMARY AND METASTATIC CANCERS OF THE LIVER: THE POTENTIAL CLINICAL VALUE FOR SPHEREX IN REGIONALIZED IMMUNOTHERAPY, HYPERTHERMIA AND RADIATION PROTECTION

George Parker and William Regelson

Departments of Surgery and Medicine and the Massey Cancer Center
Medical College of Virginia/Virginia Commonwealth University
Richmond, Virginia 23298

ABSTRACT:

Spherex is a medical device which produces controlled occlusion of arterial vessels for a half life of 15', governed by serum amylase digestion. The purposes of this study were to determine the feasibility and toxicity of the use of Spherex with standard chemotherapeutic agents for the treatment of unresectable cancers of the liver.

Twenty two patients with advanced cancer involving the liver were treated with hepatic arterial (HA) chemotherapy mixed with 900mg of 45μ biodegradable starch microspheres (Spherex-Pharmacia) in a phase II study of Spherex combined with standard chemotherapy. Thirteen patients were treated repeatedly via permanent HA catheters placed operatively using subcutaneous ports whereas 9 patients received their treatments through percutaneously femoral placed HA catheters.

Seven patients of 22 treated clearly showed regression of liver metastases with 4 showing improvement in quality of life. All but one patient studied had >50% tumor liver replacement. Twelve colorectal patients received 5FU 600mg/m^2 day 1 and mitomycin, 10mg/m^2 day 3 in 28 day cycles. Adriamycin, 30mg/m^2, was used to treat one breast cancer patient; and 2 hepatoma patients with a significant partial remission of 9 months in the patient with breast cancer. Five patients received BCNU 100-200mg; 3 with melanoma with one dramatic response that lasted 7 months; one epidermoid carcinoma responded to 2 courses of BCNU and one of thiotepa with dramatic regression and clinical improvement and is still alive at 10 months. One patient, a systemic FAM failure, has responded to 5FU/mito and 5FU alone x 3 courses. Eight patients had one course, 3:2, 3:3, 4:5, 1:5, 1:7, 1:8, 1:14 of Spherex/chemotherapy.

Mild hematologic toxicity was present in only one patient and only 1 of 22 patients had a complication with duodenitis from improper catheter placement. Transient pain in liver and nausea and vomiting were the major toxicities seen.

From our Phase II study, we feel that Spherex chemotherapy can be repeatedly administered via the HA as a convenient palliative approach to liver metastatic or primary disease. The advantages of Spherex HA occlusion combined with chemotherapy relate to the fact that 85-95% of the blood supply to metastatic carcinoma of the liver is of HA origin and Spherex achieves a tumor ischemia combined with an increase in chemotherapy concentration to the HA bed with minimal systemic toxicity.

This study demonstrates the safety and convenience of Spherex/chemotherapy combinations, but the therapeutic advantages of Spherex over other forms of hepatic arterial occlusion or chemotherapy can only be inferred unless a larger controlled or comparative study is undertaken.

Spherex occlusion has also been used experimentally in Sweden to produce transient ischemia of bowel and peripheral limbs to protect target organs from radiation injury during the occlusive period. Hyperthermia to liver metastasis can be regionally enhanced during Spherex administration and this approach deserves study.

INTRODUCTION

We now have a new approach to the treatment of liver metastases or primary hepato/biliary tumors with hepatic arterial administration of starch microspheres synthesized by Pharmacia, Uppsala, Sweden. Spherex is the clinical formulation of $45\mu \pm 5\mu$ diameter microspheres with a clinical half life of 15 minutes, governed by serum amylase degradability.

Spherex is a medical device for passively increasing local chemotherapy concentration as well as producing transient repeatable intra-tumoral ischemia. Data suggests that Spherex can provide a 3 to 11 fold increase in local chemotherapy concentration within the hepatic arterial tumor bed as amylase digests the starch releasing the occlusion and making the drug locally available (Arfors et al., 1979 a,b; Lindell et al, 1977 a,b,1978). The action of Spherex can provide an ischemic and a drug, radionucleide or monoclonal antibody localizing modality for the treatment of liver tumors.

The new and developing technology of arterial catheter placement to regional areas or via direct surgical placement, with ready access via catheters attached to subcutaneous ports, permits this technique to be of value not only in the treatment of hepatic metastases or primary hepatoma, but it will also allow for treatment in other regional areas as well, i.e., the kidney or tumors localized to the pelvic area or limbs.

Starch microspheres, unique in that they are made to be biodegradable, consist of cross-linked potato starch specially designed to become temporarily trapped at the arteriolar level. The starch undergoes digestion from the normal concentration of endohydrolases (amylase) in the serum.

Because the degree of cross-linking is greatest in the outer shell of the microspheres, the spheres maintain their shape for a considerable time during digestion. For clinical administration microspheres are provided in 1 ml normal saline (0.9%) in a concentration of 60 mg/ml.

As of March, 1985, 17 patients have received Spherex in 2 dose titration studies; 20 patients have been looked at in a blood flow (shunting study); 47 patients in 5 pharmacokinetic studies (Ensminger, 1985); 60 patients in a randomized crossover multi-center study; and there are at least 54 patients in ongoing phase II studies of patient response and convenience related to Spherex combined with standard chemotherapy (Gardner, 1985; Thulin, 1985; Aronson, 1985).

Based on observations in 22 patients, we feel Spherex offers convenience and safety as a medical device for the treatment of metastatic and primary liver tumors when combined with chemotherapy.

METHODS

Patient selection was confined to patients with unresectable liver metastases or primary hepato-biliary tumors with an estimated survival of greater than one month.

All patients had to have: Demonstrable unresectable liver metastases at the time of primary surgery; or first sign of unresectable recurrence to the liver; progressive liver metastatic growth despite systemic chemotherapy; unresectable hepatoma or biliary duct tumors; patients were at least 3 weeks post previous chemotherapy with no signs of progressive bone marrow depression.

Ineligibility included: Widespread metastatic disease significantly affecting the quality of survival; unsuitable arterial anatomy (with particular attention to avoiding shunting of chemotherapy to the stomach or duodenum); allergy or contraindication to contrast agent or chemotherapeutic agents; patients with nonmeasureable lesions on palpation, CT scan, radionuclide scan or ultrasonography; jaundice associated with signs of hepatic or renal failure; persistent bone marrow depression (WBC- <3000, platelets <90,000); significant clotting abnormalities.

Informed consent was obtained in all cases.

Preoperative tests included: Standard liver function tests (bilirubin, alkaline phosphatase, SGOT, SGPT, LDH and PTT) serum amylase, and CT scan, radionucleide scan, angiography, or ultrasound studies to delineate the size of the tumor and/or the number of metastases. CEA or alpha fetoprotein, alkaline phosphatase and other biomarker determinations were obtained as appropriate.

Selective hepatic arteriograms were obtained to determine the site of catheter tip placement before Spherex chemotherapy perfusion of the tumor bed.

Each patient had a metastatic evaluation to delineate the character of disease outside the liver prior to treatment. CBC/platelet count, renal and liver chemistries were obtained before and after each course of treatment.

Patient response was related to: Changes in measureable lesions, changes in functional status and survival.

In 9 cases, Spherex chemotherapy was given through percutaneous placed trans-femoral catheters placed by our diagnostic radiology catheter laboratory. In one case, hepatic arterial anatomy required separate catheterization to each lobe of the liver. In this patient,

(DL), failure to obtain repeated effective percutaneous catheter placement resulted in discontinuation of repeated courses.

Thirteen patients were treated through operatively placed hepatic artery catheters which were attached to a subcutaneous access portal (Port-A-Cath, Pharmacia NuTech).

Chemotherapy given with Spherex consisted of:

5 fluorouracil (5FU) 600mg/m^2 to the nearest 500mg was delivered on day 1; followed by Mitomycin C - 10mg/m^2 delivered on day 3; repeated courses were given at approximately monthly intervals.

Adriamycin 30mg/m^2 was administered at approximate 3-4 week intervals.

BCNU 100-200mg was given at 4 week intervals and Thiotepa .8mg/kg was given in 2 cases one month following previous chemotherapy with 5FU and BCNU. 5FU -600mg/m^2- was given to two patients for one course each and one patient for 3 and one for 7 courses.

The chemotherapy selected was mixed with 15 cc of Spherex suspension in our research pharmacy and transported to our radiology catheter lab. The time between preparation and Spherex/chemotherapy administration was an average of 20' with occasional patients treated one hour after preparation. The Spherex/chemo suspension was shaken just prior to administration to assure an even suspension of Spherex beads.

RESULTS

Table I shows those patients with significant regression of measurable disease. Four of these patients have had significant functional improvement with return to work in the melanoma patient (DJ) for 8 months. Following his first percutaneous treatment, he was treated via a portacath placed surgically for a total of 14 treatments and died of extra-hepatic metastases six months after his treatment was discontinued.

One breast cancer patient, (DL), required separate hepatic arterial percutaneous placement for each lobe of the liver. Despite excellent regression of liver metastases, she developed a transudative hepatic ascites (probably 2° to liver scarring) with metastases outside the liver necessitating systemic IV chemotherapy.

One hepatoma patient, (MP), had a dramatic but extremely transient regression with no improvement in survival.

Currently, we have one patient with an epidermoid cancer of unknown primary (AE) who has responded to BCNU X2 and Thiotepa X1, with dramatic regression of hepatomegaly with each course. She is currently stable after 10 months and has had no treatment for 6 months. She showed severe pain and anxiety with each procedure and hypertensive episodes related to her anxiety. Patient (GP) with breast metastases, despite liver regression to non-palpability, was lost to follow-up and died of a recurrence of previously radiated cerebral metastases.

An unknown primary patient, (BA) who previously responded to systemic FAM has shown response to 5FU/mito followed by 5FU alone at monthly intervals. Mitomycin C/Spherex produced severe nausea and vomiting and had to be discontinued.

Table 2 shows patients who were stable on treatment, without significant change in measurable disease. Two of these died of progressive intra-abdominal disease despite stabilization of their liver metastases. One patient is still stable and under study at 8+ months.

Table 3 demonstrates the severity of the extent of liver and extra hepatic disease in the patient population we treated. One colon cancer patient (NK) perforated a duodenal ulcer unrelated to the direct action of Spherex/5FU/mitomycin C. However, one colon patient (DC) developed transient duodenitis 2° to his treatment, probably related to misjudgement of catheter placement.

Significant thrombocytopenia was found in only one case (JM) after 3 courses of 5FU/mitomycin and the patient went on to 4 more courses at the same dose with no evident toxicity.

The major toxicity seen relates to pain in the liver acutely during the procedure, frequently accompanied by nausea and vomiting. Liver pain can last 1-3 days and nausea for 1-2 days. In each case pain and nausea was relieved by narcotics, sedation or antiemetics. BCNU, possibly because of its alcohol diluent, appears to produce the greatest acute pain and discomfort.

DISCUSSION

Lindell et al. (1977 a,b; 1978) first demonstrated in rats that the hepatic blood flow to the liver was temporarily reduced following intra-arterial (hepatic artery) injection of starch microspheres. Degradable starch microspheres produced a proportionally greater reduction in the blood flow to liver tumors than to healthy liver tissue.

Tuma et al. (1979) simultaneously injected tritiated actinomycin D and degradable starch microspheres into one of the renal arteries of a series of dogs. Degradation of the starch microspheres by endogenous amylase resulted in the recovery of blood flow within an hour after injection. Twenty-three per cent (23%) of the total amount of the drug was retained in the kidney one hour after the combined injection of the drug and microspheres. In contrast, when the same amount of drug was selectively injected into a renal artery by itself or was given intravenously, 17% or less of the dose injected was found in kidney tissue one hour later.

Similar regionalized effects using Spherex/ethacrynic acid injected into a renal artery was studied in dogs. Diuresis was four times greater than when the diuretic was injected intra-arterially regionally without microspheres and no effect was seen on the contralateral kidney.

These studies showed that the intra-arterial injection of a drug combined with degradable microspheres modified the systemic drug response and can produce a localized response. Previous clinical dose ranging studies, Zeissman et al. (1983) demonstrated that the most efficient method for producing transient hepatic arterial occlusion without significant back flow or lung embolization was to give 900mg of Spherex in 15 cc of physiological saline as a rapid bolus given in 1-3 minutes.

TABLE I

PATIENT AGE/SEX	DIAGNOSIS	% DISEASE	DRUGS	# COURSES	TOXICITY	RESPONSES	SURVIVAL	COMMENTS
M.J. 61/F	COLON	±50%	5FU/ Mito-C	4	nausea	partial-3mo	5 mos	
M.P. 26/M	HEPATOMA	>70%	Adria	2	N & V	tumor regress 10cm, but re-grew in 10day	6 weeks	incredible growth following initial response
D.J. 39/M	MELANOMA	>70%	BCNU		N & V liver pain	partial-70% reduc. 7 mos	13 mos	significant response
D.L. 39/F	BREAST	>70%	Adria	3	N & V	PR-9 mos	11 mos	liver controlled for 9 months
A.E. 72/F	Epidermoid unknown primary	>50%	Thiotepa BCNU/x2	3	N & V headache, pain, hyper.	liver non-palp 2X: major decrease 1X	6 mos+	mild ascites post treatment
G.P. 58/F	Unknown	>50%	5FU/ Thiotepa	1 ea.	N & V	liver decrease in size	2 mos?	CNS mets-lost to followup
B.A. 63/F	Unknown	>50%	5FU/ Mito-C x 1 5FU x 3	4	severe N & V to Mito-C	CEA decrease liver decrease	5 mos+	previous FAM respon.

TABLE II

PATIENT AGE/SEX	DIAGNOSIS	% DISEASE	DRUGS	# COURSES	TOXICITY	RESPONSES	SURVIVAL	COMMENTS
J.M. 70/M	COLON	> 70%	5FU/MitoC	8	thrombocytopenia with 4th course	stable 6 mos	7 mos	
W.L. 59/M	COLON	> 70%	5FU/MitoC	5	Fever to 103 with drugs-cycle 5	SD 3 mos then progression	12+	
A.W. 56/M	COLON	> 70%	5FU/MitoC	4	pain	SD 2 months	4 mos	progressive intra-abdom. disease-liver appeared stable
M.A. 52/F	COLON	> 70%	5FU/MitoC	4	N & V	SD 2 months	3 mos	died from progress. intraabdominal disease-liver stable
W.N. 62/M	COLON	> 70%	5FU/MitoC	4	N & V pain in liver	SD 4 months	3+ mo	
M.P. 76/F	HEPATOMA	> 50%	5FU	7	N & V	SD x 6 months	7 mos	
H.D. 63/M	HEPATOMA	50%	Adria	2		SD x 2 months	3 mos	variceal hem. severe cirrhosis

TABLE III

PATIENT AGE/SEX	DIAGNOSIS	% DISEASE	DRUGS	# COURSES	TOXICITY	RESPONSES	SURVIVAL	COMMENTS
D.C. 62/M	COLON	> 70%	5FU/MitoC	1	duodenal ulcer		2 mos	
R.H. 65/M	COLON	> 70%	5FU/MitoC	1	mild nausea		2 mos	Died of cerebral mets
R.V. 63/M	COLON	> 70%	5FU/MitoC	1	liver fail.		1 mo.	Died of liver failure
N.K. 45/F	COLON	> 70%	5FU/MitoC	1	duodenal ulcer. perf. to death		4 mos	garden variety DUD at post-no mucosal burn
R.T. 63/M	COLON	25%	5FU/MitoC	1	duodenal obst.		3 mos+	duodenal lesion not seen at time of cath. insert.
J.N. 60/F	COLON	> 70%	5FU/MitoC	3	N & V		4 mos+	prog. liver disease
J.F. 74/M	MELANOMA	> 70%	BCNU	1	N & V-mild liver pain		died-10 days later	
L.G. 17/F	MELANOMA	> 70%	BCNU	2	N & V liver pain	tumor liver necrotic at post	6 weeks	

Spherex has been used for radiation protection via transient arterial occlusion in studies by Forsberg et al (1978 a,b,c,d; 1979, 1981). Analysis of the extent of damage in relation to the irradiation dose showed a Spherex induced ischemic epithelial protective effect of a factor of about 2 for skin and intestinal radiation damage. A protective factor of about 1.5 was found for the effect on survival of the kidney. Based on this: The temporary, intra-arterial blockage and ischemia induced by degradable starch microspheres offers a new alternative for achieving appreciable protection of the skin, kidney and the intestines against tissue damage caused by x-irradiation.

Following the above preclinical studies, Aronson et al. (1979) used degradable starch microspheres, having a half-life of one hour, infused together with 5FU via a percutaneously introduced catheter into the hepatic artery. In 12 patients with liver metastases, it was demonstrated that it was possible to safely administer conventional doses of 5FU in combination with starch microspheres with minimal effects on bone marrow.

Subsequently, Dakhil et al (1981, 1982) reported on the improved regional selectivity of hepatic arterial BCNU when given in combination with Spherex. Hepatic arterial flow was reduced by 80-100% and there was a 30-90% reduction in systemic BCNU (carmustine) nitrosurea exposure.

Teder et al (1983) have shown in a comparison study of hepatic venous and peripheral venous blood that Spherex combined with Doxorubicin (adriamycin) causes a 31-72% reduction in the hepatic venous flow and a 21-42% reduction in systemic exposure to adriamycin in patients with primary or secondary liver tumors. Similarly, Gyves et al (1982,-1983) have shown that there is evidence of increased regionalized concentration of mitomycin C in conjunction with Spherex.

Spherex, because of its occlusive action combines the virtues of regional enhanced chemotherapy combined with ischemic action directed at the primary blood supply of liver metastases via the hepatic artery.

In regard to the above, it has been known that both primary and secondary liver neoplasms obtain all or almost all of their vascular supply from the hepatic artery (Breedis & Young, 1954). Effective survival of hepatic function was demonstrated by Markowitz in 1949, who showed that the dog's hepatic artery could be ligated with no undue effects provided that penicillin was given postoperatively. This was applied to the clinical control of tumor growth in 1953 and since then numerous attempts have been made to treat liver neoplasms with hepatic arterial ligation (Almersjo et al., 1972; Sivula & Sipponen, 1976; Ramming et al., 1976; Patt et al., 1983).

Although hepatic arterial occlusion produces tumor regression, the survival of the patients following hepatic arterial occlusion was not consistently longer than in an untreated control group as reported by Bengmark et al, 1974. Reports of 7-8 months of improved survival, despite liver metastases, is not dissimilar to what has been reported for bolus 5FU by direct intra-arterial infusion to liver metastases.

Following arterial occlusion, the rapid resumption of tumor growth is probably due to an early revascularization from collaterals (Bengmark & Rosengren, 1974) and "temporary dearterialization", using strangulating polyethylene slings around the hepatic artery has been tried to prevent the development of a collateral circulation (Bengmark et al., 1974). Dearterialization has been combined with local infusion of oncolytic drugs either into the portal vein (Murray-Lyon et al., 1970; Almersjo et al., 1976) or into the hepatic artery distal to the ligature (Gulessarian et al., 1972).

Patt et al., (1983) Mokka et al. (1975) have reported improved survival with hepatic arterial ligation alone and have combined it with 5FU via hepatic arterial infusion.

Ramming et al (1976) reported on hepatic ligation and chemotherapy as combined modalities also using hepatic arterial infusion. They compare their results with that of Sullivan et al. (1964) and Watkins et al. (1970) where median survivals of 15 months were reported for continuous hepatic arterial infusion without occlusion of the hepatic arteries. These results are similar to median survivals reported for systemic intravenous infusion with 5FU and are inferior to results reported by Ensminger's group (1978) with Infusaid continuous arterial Floxuridine (FUDR) infusion.

Hepatic dearterialization has been used as a technique for preparing a patient for surgical resection (Bengmark, 1974; Fortner, 1972). Sivula and Sipponen (1976) have actually performed repeated surgical dearterialization for a recurrent carcinoid with prolonged palliation.

The Spherex occlusion approach we have used is similar to the utilization of chemotherapy in conjunction with hepatic arterial ligation. However, as discussed previously, hepatic arterial ligation is associated with the rapid development of collaterals to the tumor bed (Tashiro et al., 1979) and it is hoped that the transient intermittant nature of the Spherex occlusion will preclude the development of significant collateral by pass of the hepatic artery permitting repeated chemotherapy access to the tumor bed.

Alternatively, Bengmark et al (1974), Murray-Lyons (1970), Almersjo et al. (1972) and Laufman et al. (1984) have also utilized hepatic dearterialization followed by 5FU or mitomycin infusion via the portal vein. Immediate effects were associated with decrease in tumor mass and relief of pain. In this regard, most recently the Surgical Cooperative Program (NSABP protocol C-02) (Fisher, 1984) based on a report by Taylor et al., (1977) is studying postoperative 5FU and heparin given by portal venous infusion as a 7 day adjuvant treatment for bowel cancer post surgical resection. Portal chemotherapy infusion represents an alternative or combined approach with Spherex hepatic arterial occlusion and chemotherapy.

Reviewing the regional management of liver metastases (Lee, 1983) it is our impression that for the treatment of advanced liver metastases, there has been no major advance with hepatic arterial infusion since the pioneering work of Clarkson et al., (1962), Sullivan et al. (1964) and Watkins et al. (1970). In this regard, an example of the popular continuous infusion approach has recently been reported by Daly et al. (1984) in advanced patients utilizing constant 2 week FUDR infusions. This program was accompanied by significant drug related hepatitis and duodenitis. Their therapeutic responses appear not to differ with that of the earlier results of Cady and Oberfield (1973) or Watkins et al. (1970) where 16 months of medium survival was found in responders. With hepatic tumor regression, systemic or hepatic arterial results do not differ from that observed in the randomized study of previously non-treated patients of the Central Oncology Group Study comparing hepatic arterial with intravenous chemotherapy (Grange et al. 1979).

Patt et al. (1980-81) reported on percutaneously placed hepatic arterial infusion of FUDR (floxuridine) and mitomycin for colorectal carcinoma. Of interest to Spherex occlusion, hepatic arterial thrombosis was reported to enhance the therapeutic effect of their procedure and this was supported by improved results following deliberate hepatic arterial occlusion and arterial chemotherapy for primary hepatic tumors (Patt et al., 1983).

Hatfield et al. (1982) using intra-arterial 5FU and mitomycin by continuous infusion at 6-8 week intervals reported a 90% objective partial remission rate in untreated patients as compared to 28% in previously chemotherapy treated patients. In another study involving this popular drug combination for bowel cancer, Peters et al. (1982) reported that regionalized intra-arterial infusion of mitomycin and 5FU is beneficial in patients with pelvic recurrence of colon carcinoma, despite cutaneous toxicity.

In regard to the above, while the initial enthusiasm for the work of Ensminger et al (1981) with hepatic arterial infusion via the Infusaid pump has diminished, it is the current standard by which Spherex intra-arterial chemotherapy will be judged. In this regard, median survival of 25 months from the time of diagnosis of liver metastases has been seen in patients with continuous FUDR infusion with or without mitomycin C via the Infusaid pump. In those with extra hepatic disease at the onset of treatment, median survival was limited to only 14 months (Niederhuber et al., 1984).

In our experience, with very conservative chemotherapy dosage, 7 of 22 patients showed significant regression of measurable disease (Table I). Of these, four were clearly benefitted with improvement in functional status for periods of at least 4-6 months. One patient is still stable at 9 months despite no further therapy for 6 months.

Only one colorectal cancer of 12 showed significant tumor regression, but with no dramatic effect on survival. However, 5 colorectal patients (see Table II) showed stabilization of measurable liver disease but with the numbers available, it is impossible to determine if stabilization was of real value to survival. Eight patients showed progressive disease (Table III) despite therapy and it must be stressed that all but one of our patients had 50-70% of their liver replaced by tumor, testifying to the advanced nature of the disease we treated. These results are not surprising in view of Fortner et al's. (1984) arterial chemotherapy results in colorectal liver metastases where the percent hepatic replacement was the best indicator of survival. Patients with less than 50% replacement had a 37% 2 year survival.

Bone marrow toxicity in our patients was seen in only 1 of 22 studied suggesting that dosage escalation is possible that might improve therapeutic results.

The problem of comparably evaluating our own and other experiences with Spherex chemotherapy versus that of Infusaid and other hepatic arterial approaches relates to establishing the degree of hepatic tumor involvement at the onset of study and evaluation in previously untreated patients. One of the best studies that authoratatively suggests the survival benefits of hepatic arterial chemotherapy is that of Bruckner et al. (1984) who compared treated and non-treated patients following the first sign of liver involvement (<25%) 5FU and FUDR were the drugs used and non-treated patients showed a mean survival of 7.3 (4-14 months) versus 17+ (4-46 months) in the treated groups with a remission rate of 81%.

We need to include Spherex chemotherapy in a similar controlled study where we treat patients at the first sign of recurrent liver involvement or at the time of primary colon surgery when liver metastases are found and where permanent hepatic arterial catheters with subcutaneous ports can be implanted.

Apart from chemotherapy, Spherex also provides an approach that could deliver monoclonal antibody or radionucleides in higher concentration to the hepatic tumor bed.

As discussed earlier, transient occlusion can be used to protect bowel or kidney on intra-abdominal radiation. Importantly, it suggests that a Spherex induced transient portal venous ischemia might provide a method for protecting normal liver parenchyma while radiating liver metastases supported by the intact hepatic artery. This is of interest to the renewed enthusiasm for radiating liver metastases in patients receiving Infusaid FUDR hepatic arterial infusion (Byfield et al., 1984).

Another place for Spherex would be to utilize its transient arterial occlusion to enhance heat production for the local anti-tumor action of microwave radiation to the tumor bed.

In conclusion, Spherex occlusive regionalized chemotherapy to primary and metastatic liver cancer is both safe, convenient and economical to the patient. The side effects are acceptable and responses can be impressive and meaningful to quality survival. It is apparent from our experience, with conservative chemotherapeutic dosage, that advanced colon cancer metastatic disease involving more than 50% liver replacement may not always be an acceptable group for this therapy, but Spherex chemotherapy use with smaller tumor burdens at higher or escalating dosage of chemotherapy is warranted.

(We wish to acknowledge Pharmacia U.S.A. and Uppsala, Sweden, for their support.)

REFERENCES

Almersjo, O., Gustavsson, B., Hafstrom, L., 1976, Results of regional portal infusion of 5-fluorouracil in patients with primary and secondary liver cancer, Am. Chir. et. Gynecol., 65:27-32.

Almersjo, O., Bengmark, S., Rudenstam, G.C., Hafstrom, L.O., Nilsson, L.A.V., 1972, Evaluation of hepatic dearterialization in primary and secondary cancer of the liver. Am. J. Surg.125:5.

Arfors, K-E., Forsberg, J-O., Larsson, B., Lewis, D.H., Rosengren, B., Odman, S. 1976, Temporary intestinal hypoxia induced by degradable microspheres. Nature, 262: 500.

Arfors, K-E., Tuma, R. F., Agerup, B. 1979a, Biospheres-a tool to induce transient ischemia in experimental medicine as studied in the kidney. Bib. Anat. 18:204.

Arfors, K., Aronson, K., Rothman, U., Regelson, W. 1979 b, The use of amylase biodegradable starch microspheres as a tumor infarcting and delivery system. For chemotherapy: intermittant hepatic arterial occlusion and the delivery of regionalized chemotherapy to hepatic metastases. ASCO Abst. C328/370.

Aronson, K.F., 1985, In press. Treatment of inoperable primary or secondary liver tumor by injection of mitomycin-C alone or in combination with starch microspheres (Spherex injection) into the hepatic artery.

Aronson, K.F., Hellekant, C., Holmberg, J., Rothman, U., Teder, H., 1979, Controlled blocking of hepatic artery flow with enzymatically degradable microspheres combined with oncolytic drugs. European Surgical Res. 11: 99.

Bengmark, S., Fredlund, P., Hafstrom, L. D., Vang, J., 1974, Present experiences with hepatic dearterialization in liver neoplasm. Prog. Surg. 13: 141.

Bengmark, S., Rosengren, K., 1974, Angiographic study of the liver after ligation of hepatic artery in man. Am. J. Surg. 119: 620.

Breedis, C., Young, G., 1954, The blood supply of neoplasms in the liver, Am. J. Path.30:969.

Bruckner, R., Rothmund, M., Hinterberger, R.,1984, Local infusion therapy in liver metastases of colorectal cancers. Results of a phase II study. Dtsch. Med. Wochenschr. 109:523.

Byfield, J.E., Barone, R.M., Frankel, S.S., Sharp, T.R., 1984, Treatment with combined intraarterial 5 FUDR infusion and whole liver radiation for colon carcinoma metastatic to the liver. Am. J. Clin. Oncol. 7:319.

Cady, B., Oberfield, R.A., 1974, Regional infusion chemotherapy of hepatic metastases from carcinoma of the colon. Am. J. Surg. 127: 220.

Clarkson, B., Young, C., Dierick, W., Kuehn, P., Kim, M., Berrett, A., Clapp, P., Lawrence, J.W., 1962, Effects of continuous hepatic artery infusion of antimetabolites on primary and metastatic cancer of the liver. Cancer, 15: 472.

Dakhil, S., Ensminger, W., Cho, K., Niederhuber, J., Doan, K., Wheeler, R., 1982, Improved regional selectivity of hepatic arterial bisdichloroethylnitrosourea with degradable microspheres. Cancer, 50: 631.

Dakhil, S., Ensminger, W., Cho, K., Doan, K., 1981, Improved regional selectivity of hepatic arterial Bis-dichloro-nitrosourea plus degradable starch microspheres.ASCO/AACR Abst. 22:C-194/383.

Daly, J.M., Kemeny, N., Oderman, P., Botet, J. 1984, Long term hepatic arterial infusion chemotherapy. Arch Surg. 119: 936.

Ensminger, W. D., Rosowsky, A., Raso, V., Levin, D. C., Glode, M., Come, S., Steele, G., Frei, E. III, 1978, A clinical pharmacological evaluation of hepatic arterial infusions of 5-fluoro-2-deoxy uridine and 5 fluorouracil. Cancer Res. 28: 3784.

Ensminger, W.D., 1985, A clinical pharmacological study of hepatic arterial chemoembolization with mitomycin and starch microspheres, In preparation.

Ensminger, W. D., Niederhuber, J., Dakhil, S., 1981, Totally implanted drug delivery system for hepatic arterial chemotherapy. Cancer Treat. Rep. 65: 393.

Fisher, B., 1984, Clinical trial evaluating the post operative and portal vein infusion of 5FU and Na heparin in patients with resectable adenoca of the colon. NBSAP Protocol C-02.

Forsberg, JO, 1978 a, Transient blood flow reduction induced by intraarterial injection of degradable starch microspheres. Experiments on rats. Acta Chir. Scand. 144, 275.

Forsberg, JO, Jung, B., Larsson, B., 1978 b, Radiation response modified by degradable starch microspheres. Experiment on the rat's foot. Acta Radiol. 17, 199.

Forsberg, JO, Jung, B, 1978 c, Abdominal radiation response modified by hypoxia after intra-aortal injection of starch microspheres. Experiment in the rat. Acta Radiol. 17, 353.

Forsberg, JO, Jung, B., Larsson, B., 1978 d, Mucosal protection during irradiation of exteriorized rat ileum. Acta Radiol. 17, 485.

Forsberg, JO, Joborn, H., Jung, B., 1979, Protective effect of hypoxia against radiation induced fibrosis in the rat gut. Acta Radiol. 18, 65.

Forsberg, JO, Millered, L., Graffman, S., Jung, B., Persson, E., Selen, G., 1981, Kidney radioprotection by temporary hypoxia experiments with degradable microspheres, Scand. J. Urol. Nephrol. 15:147.

Fortner, J., Nulcare, R., Solis, A., Watson, R., Golbey, R., 1972, Treatment of primary and secondary liver cancer by hepatic artery ligation and infusion chemotherapy. Ann. Surg. 178:162.

Fortner, J. G., Silva, J. S., Cox, E. B., Golbey, R. B., Gallowitz, H., MacLean, B. J., 1984, Multivariate analysis of a personal series of 247 patients with liver metastases from colorectal cancer. Ann Surg. 199:317.

Gardner, B., 1985 Personal Communication, Pharmacia, Piscataway, N.J.08854.

Grange, T. B., Vassilopoulos, P. P., Shingleton, W. W., Jubert, A. V., Elias, E. G., Aust, J. B., Moss, S. E., 1979, Results of a prospective randomized study of hepatic artery infusion with 5 fluoruracil versus intravenous 5 fluorouracil in patients with hepatic metastases from colorectal cancer. A Central Oncology Group study. Surgery 86: 550.

Gulessarian, H.P., Lawton, L. L., Condon, R. E., 1972, Hepatic artery ligation and cytotoxic infusion in treatment of liver metastases. Arch. Surg. 105: 280.

Gyves, J., Ensminger, W., Van Harken, D., Niederhuber, J., Knutsen, C., Doan, K., 1982, Improved regional selectivity of hepatic arterial mitomycin by starch microspheres. 73rd Proc. AACR 537: 127.

Gyves, J. W., 1983, Improved regional selectivity of hepatic arterial mitomycin by starch microspheres. Clin. Pharmacol. Ther. 34: 259.

Hatfield, A. K., Kammer, B. A., Danley, R. A., Miller, A. G., Jr., Houston, J. A., Harder, L., 1982, Intermittent hepatic artery perfusion for symptomatic metastatic colon carcinoma. 18th Ann. mtg. ASCO C-395.

Laufman, L. R., Nims, T. A., Guy, J. T., 1984, Hepatic artery ligation and portal vein infusion for liver metastases from colon cancer. J. Clin. Oncol. 2:1382.

Lee, Y-T. N., 1983, Regional management of liver metastases II. Can. Invest. 1: 321.

Lindell, B., Aronsen, K. F., Rothman, U., 1977 a, Repeated arterial embolization of rat livers by degradable microspheres. Eur. Surg. Res. 9: 347.

Lindell, B., Aronsen, K. F. Rothman, U., Sjogren, H. O., 1977, The circulation in liver tissue and experimental liver metastases before and after embolization of the liver artery. Res. Exp. Med. 171:63.

Lindell, B., Aronsen, K. F., Nosslin, B., Rothman, U., 1978, Studies in pharmacokinetics and tolerance of substances temporarily retained in the liver by microsphere embolization. Ann. Surgery 187: 95.

Markowitz, J., Rappaport, A., Scott, A. C., 1949, Prevention of liver necrosis following ligation of hepatic artery. Proc. Soc. Exp. Biol. Med. 70: 305.

Mokka, R. E. M., Larmi, T. K. I., Huttunen, R., Kairaluoma, M. I.,

1975, Evaluation of the ligation of the hepatic artery and regional arterial chemotherapy in the treatment of primary and secondary cancer of the liver. Ann. Chirurgae et Gynaecol. Fennae. 64: 347.

Murray-Lyon, T. M. Parsons, V. A., Blendis, T. M. Dawes, T. T., Rake, M.O., Laws, J. W. Williams, R., 1970, Treatment of secondary hepatic tumors by ligation of hepatic artery and infusion of cytotoxic drugs. Lancet 2: 172.

Niederhuber, J. E., Ensminger, W., Gyves, J., Thrall, J., Walker, S., Cozzi, E., 1984, Regional chemotherapy of colorectal cancer metastatic to the liver, Cancer 53: 1336.

Patt, Y. Z., Mavligit, G. M., Chuang, V. P., Wallace, S., Johnston, S., Benjamin, R. S., Valdivieso, M., Hersh, E. M., 1980, Percutaneous hepatic arterial infusion (HAI) of mitomycin C and floxuridine (FUDR). An effective treatment for metastatic colorectal carcinoma in the liver. Cancer 46: 261.

Patt, Y. Z., Wallace, S., Freireich, E. J., Chuang, U. P., Hersh, E. M., Mavligit, G. M., 1981, The palliative role of hepatic arterial infusion and arterial occlusion in colorectal carcinoma metastatic to the liver. Lancet 1: 349.

Patt, Y. Z., Chuang, V. P., Wallace, S., Benjamin, R. S., Fuqua, R., Mavligit, G. M., 1983, Hepatic arterial chemotherapy and occlusion for palliation of primary hepatocellular and unknown primary neoplasms in the liver. Cancer 51: 1359.

Peters, R. E., Patt, Y. Z., Chuang, V.P., Wallace, S., Fuqua, R., Mavligit, G.,1982, Palliation of pelvic recurrence of colorectal cancer by intraarterial chemotherapy. 18th Ann. ASCO, C 147.

Ramming, K. P., Sparks, F. C., Eilber, F. R., Holmes, E. C., Morton, D. C., 1976, Hepatic artery ligation and 5-fluorouracil infusion for metastatic colon carcinoma and primary hepatoma. Ann. J. Surg. 132: 236.

Sivula, A., Sipponen, P., 1976, The effect of hepatic dearterialization and redearterialization on carcinoid liver metastases. Ann. Chirurgiae et Gynaecol. Fennae. 65: 168.

Sullivan, R. D., Norcross, J. W., Watkins, E., 1964, Chemotherapy of metastatic liver cancer by prolonged hepatic-artery infusion. N. Engl.J. Med. 270: 321.

Tashiro, S., Hiraooka, T., Yoshida, M., Murata, T., Konno, T., Yokoyama, I., 1979, Combined therapy of the ligation of the hepatic artery and continuous intraarterial infusion for primary and secondary liver cancer. Chir. Gastroent. 13: 43.

Taylor, I., Brooman, P., Rowling, J. T., 1977, Adjuvant liver perfusion in colorectal cancer. Initial results of a clinical trial. Br. Med. J. 2: 1320.

Teder, H., Nilsson, B., Jonsson, K., Hellekant, C., Aspegren, K., Aronson, K.F., 1983, Hepatic arterial administration of doxorubicin (adriamycin) with or without degradable starch microspheres: a pharmacokinetic study in man.In: <u>Anthracyclines and cancer therapy</u>, Ed. H. Hansen, Excerpta Medica, Amsterdam. p. 166.

Thulin, L., 1985, Temporary reduction of hepatic arterial flow by starch microspheres (Spherex injection) in patients with hepatic tumors. In preparation.

Tuma, R. F., Forsberg, J. O., Schosser, R., Arfors, K.-E., 1979, The trapping of drugs in the microcirculation with degradable microspheres. Bibl. Anat. 210.

Watkins, E., Khazei, A. M., Nahra, K. S., 1970, Surgical basis for arterial infusion chemotherapy of disseminated carcinoma of the liver. Surg. Gynecol. Obstet. 130, 581.

Zeissman, H. A., Thrall, J. H., Gyves, J. W., Ensminger, W. D., Niederhuber, J. E., Tuscan, M., Walker, S., 1983, Quantitative hepatic arterial perfusion scintigraphy and starch microspheres in cancer chemotherapy. J. Nucl. Med. 24: 871.

SELECTIVE THERAPY OF HEPATIC CANCERS

USING MICROSPHERES

John W. Gyves

Assistant Professor of Medicine
University of Michigan
Ann Arbor, Michigan 48109

Regional chemotherapy is based on the premise that many chemotherapeutic agents display a steep dose response for toxicity and for therapeutic effect. Regional chemotherapy administration represents a means to generate increased drug exposure in the region where the tumor resides, while maintaining a lower drug exposure at the level of dose-limiting normal host tissues elsewhere in the body. Thus, even in circumstances where systemically administered chemotherapy is relatively ineffective, regional chemotherapy may improve the likelihood of response by the generation of much greater drug exposure. With sufficient regional selectivity, dose-limiting toxicity should be manifested by the normal tissues of the region infused and not by tissues elsewhere in the body. In this regard, regional chemotherapy has similarities to radiation therapy, but may be more selective in those situations where tumor and normal tissue in the treated region differ significantly in intrinsic drug sensitivity and blood supply.

Of all the forms of regional chemotherapy practiced, experience has been greatest with intra-arterial therapy and this has been most extensively applied to the treatment of primary and metastatic cancer in the liver, with a history going back more than 20 years. Pharmacokinetic analyses outlining the crucial elements and potential drug exposure increase with intra-arterial drug infusions have been carried out by Eckman et al (1), Chen and Gross (2), and Collins and Dedrick (3). Combining the advantage gained through the total body clearance (TBC) relative to regional blood flow (Q) plus that gained by regional extraction (E), the general equation defining the regional advantage of intra-arterial infusion is:

$$\text{Regional advantage} = 1 + \frac{TBC}{Q(1-E)}$$

Based on radiographic appearances using contrast angiography, it has been customary to regard tumors as being either hypo- or hyper-vascular relative to normal liver. As tumor nodules grow, the evoked new capillary bed develops at the periphery, so that the most vascularized area is the outer shell of the nodule (4). Although the central core of many tumor nodules in the liver is hypovascular, the periphery of the tumor nodules is generally hypervascular relative to normal liver, as demonstrated by nuclear tomographic scans after hepatic-arterial injection of Tc99m-macro-aggregated albumin (TcMAA) (5). The microvascular pattern is consistent with the distribution of tumor cell viability and growth, the central core of the tumor often being necrotic while the peripheral rim of actively proliferating tumor cells has an excellent blood supply. The presence or absence of a hypovascular core appears to relate more to the size of the tumor nodule than to tumor type. Nodules less than 8 cm. in diameter are uniformly hypervascular, whereas those greater than 9 cm. in diameter display a hypovascular core and a hypervascular rim as ascertained by radionuclide tomographic angiography. The density of vessels in the hypervascular regions of tumor nodules appears to be twofold to sixfold greater than in normal liver. These observations regarding regional blood flow (Q) and relative capillary density suggest a number of methods whereby tumor hypervascularity can be used for selective therapeutic advantage.

Microspheres of 40-80 μm diameter, when injected as a homogeneous suspension into the hepatic artery, should lodge in the hepatic-arterial microvasculature in direct proportion to regional blood flow throughout that watershed (6). As mentioned above, the hepatic-arterial injection of TcMAA with nuclear tomography provides a means to determine the relative blood flow distribution between normal liver and tumor nodules within liver and, thus, to monitor selective delivery of therapeutic microspheres to tumor (5,6). One method for decreasing hepatic-arterial blood flow is the use of microparticulates or microspheres. At sufficiently high doses, approximating 90 million biodegradable starch microspheres (40-μm diameter, Pharmacia, Uppsala, Sweden), hepatic-arterial blood flow can be totally blocked in about 25% of patients (7). By 30 minutes after hepatic-arterial tree as ascertained by contrast angiograms. In the remaining 75% of patients, hepatic-arterial flow decreases by 80% and arterial-venous shunting occurs. Thus, the use of the hepatic-arterial starch microspheres may be an additional method to deliver more drug to tumors within the liver.

The concurrent hepatic-arterial injection of a suspension of starch microspheres in a drug solution has the potential of temporarily holding the drug solution in the hepatic-arterial capillary bed, thus allowing more time for the higher drug concentration to move into surrounding tissue. Carmustine and mitomycin have been examined in conjunction with starch microspheres given via the hepatic artery (7,8). These agents were chosen due to their rapid tissue uptake and mechanism of action as alkylating agents. Because of increased drug delivery to the liver and hepatic tumor, systemic drug exposure was reduced up to 90% for carmustine (7) and 70% for mitomycin (8) when the drug was given with starch microspheres versus drug injection alone. In hypervascular regions of tumors, more drug solution should be held up as compared to the drug entrapment in less vascular regions of normal liver. As the microspheres are digested, their diameter progressively decreases (from 40 μm initially) and the drug column moves distally into the capillary bed. Due to complete dissolution of the starch microspheres, subsequent doses can be administered without destroying access to the tumor microcirculation. These studies have prompted the initiation of phase II clinical trials of hepatic-arterial carmustine and mitomycin with starch microspheres.

REFERENCES

1. W. W. Eckman, C. S. Patlak, and J. D. Fenstermacher, A critical evaluation of principles governing the advantages of intra-arterial infusions, J. Pharmacokinet Biopharm 2:257-285, (1974).
2. H. S. G. Chen, and J. F. Gross, Intra-arterial infusion of anti-cancer drugs: Theoretic aspects of drug delivery and review

of responses, <u>Cancer Treat Rep</u> 64:31-40, (1980).
3. J. M. Collins, and R. L. Dedrick, Pharmacokinetics of anticancer drugs, <u>in</u>: Pharmacologic Principles of Cancer Treatment, Chabner B, ed., W. B. Saunders Co., Philadelphia, (1982).
4. B. A. Warren, The vascular morphology of tumors, <u>in</u>: Tumor Blood Circulation: Angiogenesis, Vascular Morphology and Blood Flow of Experimental and Human Tumors, Peterson HI, ed., CRC Press, Inc., Florida, (1979).
5. J. W. Gyves, H. A. Ziessman, W. D. Ensminger, J. H. Thrall, J. E. Niederhuber, J. W. Keyes, Jr., and S. Walker, Definition of hepatic tumor microcirculation (SPECT), <u>J Nucl Med</u>. 25: 972-977 (1984).
6. H. A. Ziessman, J. H. Thrall, J. W. Gyves, W. D. Ensminger, J. E. Niederhuber, M. Tuscan and S. Walker, Quantitative hepatic arterial perfusion scintigraphy and starch microspheres in cancer chemotherapy. <u>J Nucl Med</u> 24:871-875 (1983).
7. S. Dakhil, W. D. Ensminger, K. Cho, J. Niederhuber, K. Doan, and R. Wheeler, Improved regional selectivity of hepatic arterial BCNU with degradable microspheres. <u>Cancer</u> 50:631-635 (1982).
8. J. W. Gyves, W. D. Ensminger, D. Vanharken, J. Niederhuber, P. Stetson, S. Walker, Improved regional selectivity of hepatic arterial mitomycin by starch microspheres, <u>Clin Pharmacol Ther</u> 34:259-265 (1983).

SECTION I: PROTRACTED ADMINISTRATION OF ANTINEOPLASTIC CHEMOTHERAPY
            AGENTS

       B. Clinical Studies

PRELIMINARY RESULTS OF A RANDOMIZED STUDY OF
INTRAHEPATIC INFUSION VERSUS SYSTEMIC INFUSION
OF FUDR FOR METASTATIC COLORECTAL CARCINOMA

Nancy Kemeny[1] and John Daly[2]
[1]Associate Attending Physician, Solid Tumor Service
Department of Medicine; [2]Associate Attending Surgeon
Colo-Rectal Service, Department of Surgery
Memorial Sloan-Kettering Cancer Center
1275 York Ave., New York, N.Y. 10021

Development of a totally implantable infusion pump produced a renewed interest in hepatic infusional therapy. An initial study employing this pump and continuous hepatic infusion of FUDR produced an 83% response rate.[1] However, further work with this method brought the mean response rate down to 59%[2] which, however, is still higher than the mean response rate obtained with systemic chemotherapy.

Although surgical and technical complications with the use of an implantable pump have been minimal, chemotherapy-related complications have been substantial. In the study at Memorial Sloan-Kettering Cancer Center (MSKCC) on 45 patients using the dose of .3mg/kg/day of FUDR for 2 weeks alternating with 2 weeks of saline, 30% had endoscopically documented gastrointestinal ulceration. If severe gastritis and duodenites are included, 50% had significant gastrointestinal disease. Hepatic toxicity was also seen frequently; an elevation of serum bilirubin level above 3 mg/dl was observed in 20% of the patients, and an elevation of SGOT level 3 times above baseline values in 46%. Two patients developed strictures of their bile duct resembling sclerosing cholangitis[2].

The most useful laboratory test to help monitor and avoid some of these side effects is glutamic oxalacetic transaminase (SGOT). If one were to obtain SGOT values every 4 weeks, these spikes at 2 weeks at the end of the infusion would be missed. In re-evaluating our first group of patients, we noted that this pattern of SGOT elevation had occurred earlier in 11 out of 12 patients with ulcer disease and in all patients who had an increase in bilirubin.

In order to see the true impact of hepatic infusion therapy on response and survival, one needs a prospective randomized study with stratification for important parameters such as liver involvement and certain indicative laboratory values.

Many investigators have shown the extent of liver involvement by tumor is an independent prognostic factor influencing survival[3,4,5]. In evaluation of our own patient population the median survival was 22 months for patients with less than 20% involvement and only 6 months for those with greater than 60% involvement.

The influence of certain laboratory parameters on response and survival was evaluated in a study at MSKCC in 200 patients with advanced metastatic colorectal carcinoma. The most significant factor affecting both response and survival was the initial lactic dehydrogenase (LDH) level. Patients whose initial LDH and carcinoembryonic antigen (CEA) levels were normal had a median survival of 32 months versus only 8 months for those who originally had abnormal values of LDH and CEA.[6]

This prospective randomized study will compare intrahepatic infusion to systemic infusion applying the same chemotherapeutic agent (FUDR), schedule, and method of administration. Patients with measurable metastatic colorectal carcinoma to the liver, without extrahepatic disease, are eligible. Patients with a Karnofsky Performance Status (PS) less than 60% and a serum bilirubin greater than 4.0 mg/dl are excluded.

After stratification by LDH level (<300 U/l vs ≥ 300 U/l) and percent liver involvement ($< 50\%$ vs $\geq 50\%$), all patients are randomized to either intrahepatic or systemic therapy prior to surgery. The extent of liver involvement is assessed medically by evaluating computerized tomography (CTT) and/or radionuclide liver scans. The information is placed in a sealed envelope and opened only after the extent of liver involvement is assessed by a surgeon during exploratory laparotomy. If there is a disagreement between preoperative and surgical assessment of the extent of liver involvement, the patient is re-randomized.

All patients undergo exploratory laparotomy not only for the placement of a hepatic artery catheter and Infusaid pump but also to ensure that two arms of the study are comparable, by accurately defining the extent of liver involvement and assuring that there is no disease outside the liver. Any patient with a resectable hepatic lesion or extrahepatic disease is considered ineligible for the protocol.

Patients randomized to intrahepatic therapy have the hepatic artery catheter connected to the pump. In the systemic group, the hepatic catheter is connected to an infus-a-port, and the pump is connected to an additional catheter placed in the cephalic vein (Fig. 1). If the disease progressed in a patient in the latter group, a minor surgical procedure would allow a crossover to intrahepatic therapy (Fig. 2), thereby also allowing further evaluation of the efficacy of regional therapy.

The drug, FUDR, is administered by continuous infusion for 14 days via an Infusaid pump in both groups. However, the starting dose is 0.3 mg/kg/day for the intrahepatic group and 0.125 mg/kg/day for those receiving systemic infusion.

One hundred and six patients have been referred for entry into the study. Seven patients refused randomization, and 3 were excluded because of anomalous arterial blood supply, i.e., more than 3 vessels perfused the liver. Therefore, 96 patients were randomized preoperatively (Table 1).

Thirty-seven patients were excluded from the study after surgical exploration for the following reasons: resectable disease in 16 patients, extra-hepatic disease in 17, no tumor in 3, and intra-abdominal infection in one. Fifty-nine patients, therefore, have had the pump placed in the randomized study. The two groups were comparable (Table 2); they were well matched with respect to percent liver involvement, initial laboratory values, PS, age, and unfavorable prognostic factors (Table 3). Sex was the only parameter by which they were not well matched; there were only 7/30 females in the intrahepatic group, while 12/29 patients were women in the systemic group. When the data was analyzed to see if gender influenced response rate, this appeared not to be the case.

Table 1. INTRAHEPATIC VS. SYSTEMIC FUDR INFUSION

REFERRED		106
Refused randomization	7	
Two or more arterial supply	3	
RANDOMIZED		96
Excluded		37
Resected	16	
Extrahepatic disease	17	
Infection	1	
No tumor	3	
ENTERED		59

Table 2. INTRAHEPATIC VS. SYSTEMIC FUDR INFUSION: PATIENT CHARACTERISTICS

	Intrahepatic (n=30)	Systemic (n=29)
Age (yrs)*	61	62
KPS (%)*	80	80
% Liver involvement	40	45
Sex (M/F)	23/7	17/12

*median

Table 3. INTRAHEPATIC VS. SYSTEMIC FUDR INFUSION: INITIAL LABORATORY DATA

		Intrahepatic (n=30)	Systemic (n=29)
LDH	> 500 U/l	8	13
CEA	> 100 ng/ml	12	16
WBC	> 10,000 cells/mm^3	4	6
Albumin	< 4.0 gm/dl	10	9
Alk Phos	> 300 U/l	10	10

Responses are assessed in the following manner. A complete response denotes the disappearance of all evidence of disease by CTT, liver scan, physical examination and CEA levels. A partial response (PR) is defined as a greater than 50% reduction in the size of measurable disease by CTT or liver scan. If the liver is palpable, a 50% reduction in liver measurements is also required. A minor response (MR) is 25-50% reduction in the size of measurable disease. CEA reductions are noted but are not used to define a response.

To date, 11 PR's in 26 evaluable patients were seen in the intrahepatic group and 8 PR's in 24 evaluable patients in the systemic group. The median duration of response is 7 months for both groups. There were two MR's in each group. In the systemic group, 3 patients have stable disease for 4-13+ months. Fifteen patients in the intrahepatic group and 10 patients in the systemic group have had more than 50% reduction in CEA level. (Table 4).

In evaluating the patients who had a crossover, there seems to be some relationship between initial response and the chance of getting a response from the crossover (Table 5). Five patients who responded and then failed systemic treatment have been crossed-over to the intrahepatic treatment. Three patients responded and one had a transient improvement which was followed by thrombosis of the hepatic artery and progression of disease. Six of 7 patients who originally failed systemic infusion also failed intrahepatic infusion (Table 5).

Table 4. INTRAHEPATIC VS. SYSTEMIC FUDR INFUSION

	Intrahepatic	Systemic
TOTAL ENTERED	30	29
too early	3	2
Inadequate trial	1	3
EVALUABLE	26	24
Partial Response	11	8
Minor Response	2	2
Stable	3	3
Reduction of CEA (>50%)	15	10

Table 5. RESULTS OF CROSSOVER FROM SYSTEMIC TO INTRAHEPATIC FUDR INFUSION

	N=.12			
Response to Systemic FUDR	PR (5)		NR (7)	
Response to Intrahepatic FUDR	PR (3)	NR (2*)	NR (6)	Too early (1)

The toxicity has been quite different between the two groups (Table 6). In the intrahepatic group, the toxicity has been mainly gastrointestinal and hepatic. Four of 30 patients developed significant gastrointestinal ulcers documented by endoscopy, and a fifth patient had severe gastritis. Seventeen patients developed an elevation of SGOT greater than 100% over baseline value, and 6 developed a significant elevation of serum bilirubin (as high as 10.0mg/dl in one patient). Three of the 6 went on to develop biliary sclerosis. In the systemic group, the major toxicity has been diarrhea, seen in twenty-two patients. In 3 patients, sigmoidoscopy revealed sigmoid ulcerations suggestive of colitis. Both patients required hospitalization for supportive care including IV hydration.

Table 6. INTRAHEPATIC VS. SYSTEMIC FUDR INFUSION: TOXICITY IN EVALUABLE PATIENTS

	Intrahepatic (n=30)	Systemic (n=29)
Ulcer	4	0
Diffuse gastritis	1	1
SGOT > 2 x baseline	17	1
Bilirubin > 3.0 mg/dl	6	0
Diarrhea	1	22
Colitis	0	3

Although the starting dose for the intrahepatic therapy was twice as high as the systemic therapy, doses in both arms were quite similar after the third cycle of treatment. (Table 7 and 8). In the systemic arm, the starting dose was 0.15 mg/Kg/day x 14 days in the first 9 patients. After 2 patients developed severe diarrhea, the dose was reduced to 0.125 mg/Kg/day.

Table 7. DOSE ADJUSTMENTS FOR INTRAHEPATIC CHEMOTHERAPY N=30

	No. Pts.
REDUCTION AFTER 1st TREATMENT*	6
2nd TREATMENT	8
3rd TREATMENT	4
MEDIAN DOSE AFTER 3rd TREATMENT 0.2 mg/kg/day	

*Starting dose 0.3 mg/kg/day

Table 8. DOSE ADJUSTMENT FOR SYSTEMIC CHEMOTHERAPY N=29

	No. Pts.	Reduction	Escalation
Original Dose FUDR 0.15 mg/kd/day	9	5 pts.	--
Subsequent Starting Dose FUDR 0.125 mg/kg/day	16	1	9 pts.

Median dose after 3rd treatment 0.15 mg/kg/day

Another significant difference between the 2 groups was the development of extrahepatic disease. In the intrahepatic group, 15 patients have already developed extrahepatic disease (9 lung, 4 intra-abdominal and 4 bone). At the present time, 6 patients in the systemic group have developed extrahepatic disease (Table 9).

Table 9. INTRAHEPATIC VS. SYSTEMIC FUDR INFUSION: DEVELOPMENT OF EXTRAHEPATIC DISEASE

	Intrahepatic (n=30)	Systemic (n=29)	Crossover (n=12)
Lung	9	2	3
Abdomen	4	0	0
Bone	4	1	0
Pelvis	0	3	0
Adrenal	1	0	0
Spinal	1	0	0
Total # Pts.	15*	6	3

There has been no difference in survival between the two treatment groups. The median survival, at the present time, for the intrahepatic group is 11 months, while it has not been reached for the systemic group because of limited follow-up thus far and very few deaths.

CONCLUSIONS

Although it is still too early to reach definite conclusions, the following observations on the infusional chemotherapy in hepatic metastases from colorectal carcinoma can be made:
1) The development of extrahepatic disease is more common with intrahepatic infusion than with systemic infusion.
2) Gastrointestinal toxicity is common with both types of infusion; upper gastrointestinal ulceration is observed with intrahepatic infusion, whereas diarrhea is observed with systemic infusion. Hepatic toxicity is seen with intrahepatic therapy but not with systemic therapy.
3) Response rates for the intrahepatic and systemic groups (42% and 33% respectively) are similar, although it is premature to draw any conclusions.

REFERENCES

1. Ensminger W., Niederhuber J., Gyves J. Thrall J., Cozzi E., Coan: Effective control of liver metastases from colon cancer with an implanted system for hepatic arterial chemotherapy. Proc ASCO 1:94, 1982.
2. Kemeny N., Daly J., Oderman P., Shike M., Chun H., Petroni G., Geller N.: Hepatic artery pump infusion: Toxicity and results in patients with metastatic colorectal carcinoma. J Clin Oncol 2:595-600, 1984.
3. Wood C.B., Gillis C.R., Blumgart L.H.: A retrospective study of the natural history of patients with liver metastases from colorectal cancer. Clin Oncol 2:285-288, 1976.
4. Pettavel J., Morganthaler F.: Protracted arterial chemotherapy of liver tumors: An experience of 107 cases over a 12-year period. In: Ariel IM (ed) Progress in Clinical Cancer. Grune & Stratton, New York, 1978, pp 217-233.
5. Nielsen J., Balslev I., Jensen H.E.: Carcinoma of the colon with liver metastases. Acta Chir Scand 137:463-465, 1971.
6. Kemeny N., Braun D.W.: Prognostic factors in advanced colorectal carcinoma. Importance of lactic dehydrogenase level, performance status and white blood cell count. Am J Med 74:786-798, 1983.

COPBLAM: INFUSION CHEMOTHERAPY FOR LARGE CELL LYMPHOMA

Morton Coleman, D. Barry Boyd, Bernard Bernhardt, Gary Gerstein, and Samuel Kopel

Oncology Service, Division of Hematology-Oncology
The New York Hospital & the Dept. of Medicine, Cornell University Medical College, N.Y., N.Y. 10021

Early in 1977, a new combination chemotherapy program was initiated at the New York Hospital-Cornell Medical Center for large cell lymphoma (LCL). This program, known as COPBLAM: {cyclophosphamide, Oncovin, (vincristine), prednisone, bleomycin, Adriamycin, Matulane (procarbazine)} was an intensive multidrug regimen designed to maximize tumor cell kill[1]. Unique to this treatment for LCL was the incorporation of then novel concepts and features which were as follows:

1) Dosage escalation provisions for two major drug components (cyclophosphamide, Adriamycin) appropriate to patient tolerance, thereby allowing fullest implementation of these agents. Protocols in the past had provided dosage reduction schedules when treatment proved too intense, but few had ever contained provisions for increasing treatment intensity, notwithstanding the steep dose-response relationship of these drugs[2].

2) Treatment cycles of 21 days, in contrast to the customary monthly schedules used in the past. Toxicity from both cyclophosphamide and Adriamycin is usually ameliorated in this interval, and shorter cycles provided for further intensification of treatment[2].

3) The use of six drugs rather than the customary four or five, all putatively non-cross-resistant with differing mechanisms of action. Procarbazine was added to the standard BACOP (bleomycin, Adriamycin, cyclophosphamide, Oncovin, prednisone) combination because of its ability to cross the blood brain barrier. Its previous incorporation in the COP (cyclophosphamide, Oncovin, prednisone) regimen to form the C-MOPP (cyclophosphamide, Oncovin, procarbazine, prednisone) program greatly augmented the cure rate[3].

4) The fuller use of nonmyelosuppressive agents, particularly vincristine, prednisone and bleomycin, of which the latter was given at day 14, allowing further tumor treatment in the face of nadir blood counts.

THE COPBLAM schedule, as outlined in Figure 1 was designed specifically for ease of outpatient administration and to be completed in a brief six months. Thereafter, patients were thoroughly reevaluated for com-

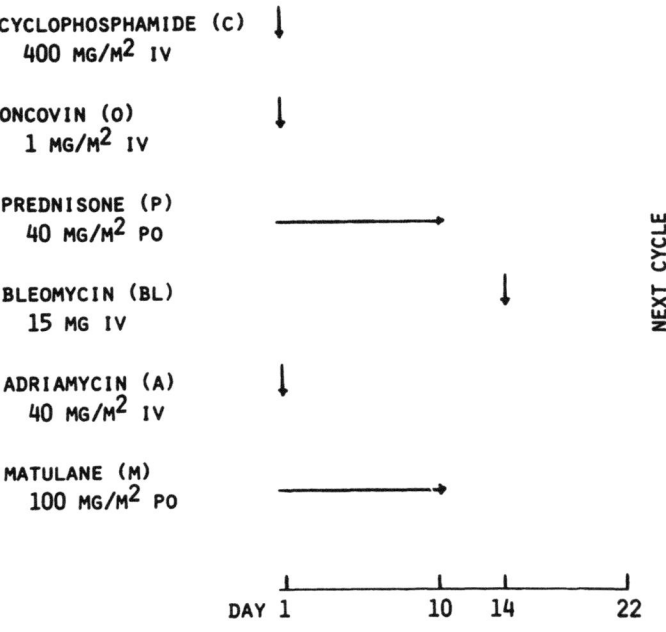

Figure 1. COPBLAM schedule

pleteness of response and, if found free of disease, all treatment was discontinued; no maintenance therapy was given[1].

Of 48 patients studied, 33 had not received prior treatment with chemotherapy. Among such previously untreated patients, 73% achieved a complete response. Among the complete responders, there were two unrelated deaths and four relapses, all of whom have subsequently died. The treatment was fully successful in 60% of all patients. At a minimum followup time off treatment in excess of 4 years, 55% remain alive and free of disease with the survival curve remaining at a plateau beyond the first two years. Most patients with large cell lymphoma, if they are to relapse, do so within two years, since the rapid growth rate and cellular kinetics promotes early relapse[2,4,5]. Since the majority of patients have survived the disease and are ostensibly cured by COPBLAM treatment, the use of the median survival as a measure of treatment outcome has been rendered obsolete by the regimen and the percentage of patients within the survival plateau becomes the benchmark. The therapeutic efficiency of COPBLAM has also been confirmed by Armitage and associates who, using a modified version of this protocol, have achieved similar results in a larger group of lymphoma patients[6].

Despite these excellent results, a significant cohort of LCL patients fail or only partially respond to treatment such as COPBLAM. Recently, continuous infusion chemotherapy has been advocated to enhance therapeutic efficacy of conventional agents[7]. This has been based on the rationale that, in heterogenous tumor cell populations, continuous exposure to cell-cycle specific agents increases the fraction of cells exposed and killed during that sensitive phase. In addition, prolonged exposure may increase intracellular concentrations of drugs and thus cell kill in cell populations having heterogeneous rates of transport across cell membranes[7,8]. Bleomycin is a cycle-specific (G2-M) agent with a serum half life of under two hours[9]. Vincristine, a known mitotic inhibitor, has a similarly short half life of less than four hours[10]. Additionally, both agents are relatively nonmyelosuppressive, making them suitable for continuous infusion therapy with minimal risk of excess bone marrow toxicity.

Both in vivo and in vitro data support the potential utility of continuous infusion bleomycin and vincristine. In the mouse Lewis lung carcinoma implant model[11], bleomycin infusion produced more dramatic tumor growth inhibition, without associated pulmonary toxicity, than two bolus regimens of equal total dose. In human lymphoma cell lines, in vitro, increasing exposure time to bleomycin from four to 35 hours resulted in a marked reduction in cell survival[12]. In Phase I trials of infusional vincristine, Jackson[13,14] noted a correlation between drug concentration and exposure time necessary for in vitro tumor cell kill and levels sustained by continuous infusion in vivo. Objective responses were obtained in 37% of these patients with resistant malignancies, including 4 of 6 with non-Hodgkins lymphoma. Several nonrandomized trials in patients with advanced malignancies, including cervical and testicular carcinoma[15], and Hodgkins and non-Hodgkins lymphoma[16,17], many of whom were refractory to conventional bolus bleomycin, have shown responses to infusional bleomycin. In addition, several studies have suggested that pulmonary toxicity on infusional regimens is less than with bolus dosing[11,18].

In view of these theoretical advantages, a nonrandomized trial of continuous infusion vincristine and bleomycin, with oral prednisone and followed by methotrexate in responders, was undertaken at the New York Hospital-Cornell Medical Center in patients with resistant non-Hodgkins Lymphoma[19].

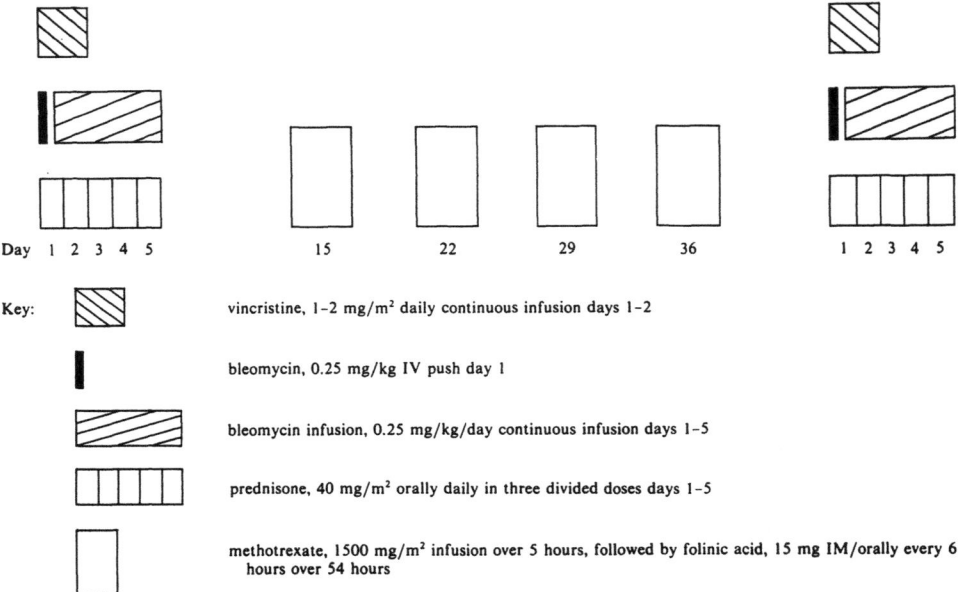

Figure 2. Schema of vincristine-bleomycin infusional treatment for resistant lymphoma.

As outlined in Figure 2, patients received vincristine 2 mg/m^2 daily by continuous IV infusion on days 1 and 2, bleomycin 0.25 mg/kg IV bolus day 1, followed by bleomycin 0.25 mg/kg by continuous IV infusion on days 1-5, prednisone 40 mg/m^2/day daily on days 1-5, and, in patients showing response by day 15 to the above regimen, methotrexate 1500 mg/m^2 by infusion over five hours followed by leukovorin rescue, on days 15,22,29 and 36. This regimen was given at six week intervals until relapse or progression occurred. In sixteen heavily pretreated patients using this infusional program, all of whom demonstrated progressive disease on intravenous bolus vincristine and bleomycin, half responded, with 3/16 (19%) achieving a complete remission. Although responses were short-lived and the doses of bleomycin larger than those employed in the initial bolus treatment, the mode of administration clearly appeared to play a large role in the response. Toxicity of the regimen was quite tolerable with stomatitis representing the major side effect (Table 1, see appendix). Of note, significant bleomycin pulmonary toxicity was not encountered, a reflection perhaps of the short treatment time owing to the generally poor prognosis of these patients.

Given the excellent results achieved with COPBLAM and the apparent improved therapeutic index of infusional bleomycin and vincristine, COPBLAM III was introduced, melding these two approaches to lymphoma treatment[20]. A COPBLAM II regimen, consisting of subcutaneous bleomycin administration on an outpatient basis, was aborted when patient compliance with self-administration proved problematical. Infusional treatment in COPBLAM III assured accurate administration and delivery. The infusions were given, however, in alternating cycles with bolus treatment. In an effort to abrogate pulmonary toxicity, the dose of bleomycin was reduced by one third, compared to the dose used in the pilot study of resistant lymphoma. COPBLAM III consisted of daily infusional Oncovin 1.0 mg/m^2 days 1 and 2, bleomycin 7.5 mg/IV push day 1, followed by daily infusions of 7.5 mg/m^2 days 1-5, prednisone 40 mg/m^2 po days 1-5, Matulane 100 mg/m^2 po days 1-5, Adriamycin 35 mg/m^2 IV push day 1 and cyclophosphamide 350 mg/m^2 IV push day 1, given every six weeks for six cycles. Between each of these inpatient cycles, every third week, as outpatients, patients received bolus Oncovin 1.0 mg/m^2, and the remaining medications as inpatient cycles, but with bleomycin omitted. Similar to COPBLAM, doses of cyclophosphamide and Adriamycin were escalated; however, both the starting dose as well as the escalations were smaller since infusional treatment in pilot studies appeared to augment both the hematologic toxicity as well as the tumor activity of the two major myelotoxic drugs. The starting doses thus were 70% of those in COPBLAM and the escalation was half, 50 mg. and 5 mg. respectively, per treatment.

Of 54 patients studied using COPBLAM III, 43 were evaluable. Complete responses were obtained in 86% (37/43) and a plateau phase (patients alive and free of disease) was achieved in 70% (30/43) of patients[20]. There were four toxic deaths, 3 relapses, 4 partial responses and 2 nonresponses. Median and mean time of followup was 22 months. Median time to complete remission was 2 cycles. Bleomycin toxicity occurred in 18 patients and 2 proved fatal. Median time to bleomycin toxicity was 4 cycles (table 2).

Table 2

	COPBLAM (%)	COPBLAM III (%)
Complete Response	73	86
Survival Plateau (pts. alive, well and off treatment)	55	70
Potential Cure	60	79

While the results with COPBLAM III were superior to COPBLAM (Table 2) considerably more expense, hospitalization and time were expended to achieve these results. COPBLAM III required three more months of treatment and six hospitalizations. Were a 15% improvement in complete remissions, a 15% improvement in overall survival plateau and a 19% improvement in potential cure rates (excluding inadvertent deaths, either treatment or nontreatment related) worth the expenditure of time and effort, particularly since comparisons are being made between sequential rather than randomized, concomitant studies? We think so. Although both regimens had a rather unfavorable patient mix from a prognostic standpoint, it is our impression that those patients treated with COPBLAM III were particularly challenging. We are currently evaluating various prognostic factors between the two regimens to determine if our impression is substantiated. If true, COPBLAM III may prove particularly useful in those patients with aggressive disease and multiple poor prognostic factors, reserving the more easily administered and better tolerated outpatient COPBLAM for better prognosis patients.

If the toxic deaths could in some manner have been abrogated in COPBLAM III, the potential for cure would approach 80%, a noteworthy achievement. The major limiting toxicity, occurring in 42% of patients, and resulting in 2 of 4 toxicity deaths was due to bleomycin. This occurred at a median of four cycles. In contrast, complete remissions were attained in a median of two cycles, confirming findings by others that patients who will respond do so early in treatment[4,6]. Patients failing to attain a complete remission on COPBLAM III, did so generally by developing progressive disease during the outpatient bolus treatment cycle. Early relapses during treatment also occurred almost exclusively during this interval between infusional therapy. By intensifying treatment further, by utilizing multiple, sequential rather than alternate cycle infusional treatment, and by reducing the total number of bleomycin infusions given we hope to further improve treatment results while reducing toxicity. The outlook for infusional treatment of large cell lymphoma is indeed promising.

REFERENCES

1. Laurence, J., Coleman, M., Allen, S.L., Silver, R.T., Pasmantier M. Combination chemotherapy of advanced diffuse histiocytic lymphoma with the six-drug COP-BLAM regiman. Ann. Intern. Med. 1982, 97: 190-5.
2. Coleman, M. Chemotherapy of large cell lymphoma: optimism and caution. Ann. Intern. Med. 1985; 102:140-2.
3. DeVita, V.T., Jr., Canellos, G.P., Chabner, B., Schein, P., Hubbard, S.P., Young, R.C. Advanced diffuse histiocytic lymphoma, a potentially curable disease: results in combination therapy. Lancet. 1975; 1:248-50.
4. Klimo, P., Connors, J.M. MACOP-B chemotherapy for the treatment of advanced diffuse large-cell lymphoma. Ann. Intern. Med. 1985; 102: 596-602.
5. Frei, E., III. Progress against lymphoma. In: Holand, J.F., Thomas, E.D., Freeman, A.I., Frei, E., III, Freireich, E.J., Arlin, A.Z., eds. Progress and Controversies in Leukemia and Lymphoma. New York: Audio Visual Medical Marketing, Inc.; 1983:40-47.
6. Armitage, J., Hutchins, M., Moravec, D., et al. Intensive 6-drug combination chemotherapy for diffuse aggressive lymphoma (Abstract). Proc. Am. Soc. Clin. Oncol. 1984; 2:1289-1304.
7. Vogelzang, N.J. Continuous infusion chemotherapy: A critical review. J. Clin. Oncol. 1984; 2:1289-1304.
8. Carlson, R.W., Sikic, B.I. Continuous infusion or bolus injection in cancer chemotherapy. Ann. Intern. Med. 1983; 99:823-833.
9. Broughton, A., Strong, J.E., Holoye, P.Y., et al. Clinical pharmacology of bleomycin following intravenous infusion as determined by radioimmunoassay. Cancer 1977; 40:2772-2778.
10. Jackson, D.V., Jr., Bender, R.A. The clinical pharmacology of the vinca alkaloids, epipodophyllotoxins and maytansine. In: Pindedo, H.M., ed. Clinical Pharmacology of Antineoplastic Drugs. Elsevier/North Holland Biomedical Press; 1978: 277-294.
11. Sikic, B.I., Collins, J.M., Mimnaugh, E.G. and Gram, T.E. Improved therapeutic index of bleomycin when administered by continuous infusion in mice. Cancer Treat Rep. 1978; 62:2011-2017.
12. Drwinko, B., Novak, J.K. and Barranico, S.C. The response of human lymphoma cells in vitro to bleomycin and 1,3-bis (2-chldroethyl)-1-Nitrosourea. Cancer Res. 1972; 32:1206-1208.
13. Jackson, D.V., Jr., Bender, R.A. Cytotoxic thresholds of vincristine in a murine and human leukemia cell line in vitro. Cancer Res. 1979; 39:4346-4349.
14. Jackson, D.V., Jr., Sethi, S., Spurr, C.L., et al. Intravenous vincristine infusion: Phase I Trial. Cancer 1981; 38:2559-2564.
15. Krakoff, I.H., Cvitkovic, E., Currie, V., et.al. Clinical pharmacologic and therapeutic studies of bleomycin given by continuous infusion. Cancer, 1977; 40:2027-2037.
16. Hollister, D., Silver, R.T., Gordon, B., Coleman, M. A new technique for treating resistant lymphoma: Vincristine-Bleomycin infusion with high dose methotrexate (Abstract). Proc. Am. Soc. Clin. Oncol. 1980; 21:368.
17. Ginsberg, S.J., Crooke, S.T., Bloomfield, C.D., et al. Cyclophosphamide, doxorubicin, vincristine and low dose continuous infusion

bleomycin in nonHodgkins Lymphoma: Cancer and Leukemia Group B Study #7804. Cancer 1982; 49:1346-1352.
18. Cooper, K.R. and Hong, W.K. Prospective study of pulmonary toxicity of continuously infused bleomycin. Cancer Treat Rep. 1981; 65:419-425.
19. Hollister, D., Silver, R.T., Gordon, B., Coleman, M. Continuous infusion vincristine and bleomycin with high dose methotrexate for resistant nonHodgkins Lymphoma. Cancer 1982; 50:1690-1694.
20. Boyd, D., Coleman, M., Adeler, K., et al Six drug polychemotherapy utilizing infusional Bleomycin and Vincristine for diffuse large cell lymphoma (Abstract). Blood, 1984; 64(5):179a.

APPENDIX

Table 1. Side effects of COPBLAM regimen.

Patient	Stomatitis	Dermatitis	Fever	Leukopenia	Thrombo-cytopenia	Pares-thesias	Myalgias	Arrythmia	Nausea	Ileus	Pneumonitis	No toxicity
1	++							+				
2		+		+		+						+
3	+											
4	++	+	+	++	+							
5	++		++	++		+						
6	++		++		+		++					
7			+		+				++			
8								+				
9												++
10												++
11				(+)								
12	+			+								
13	++			+								
14	+									++		
15	+		+	+							+	
16												
Totals	10/16	2/16	5/16	7/16	3/16	2/16	1/16	2/16	1/16	1/16	1/16	4/16
%	63	13	31	44	19	13	6	13	6	6	6	25

+: Finding present within three weeks of infusion.
++: Severity of toxicity required reduction of infusion.
( ): Toxicity present before beginning treatment.

ADRIAMYCIN CONTINUOUS I.V. INFUSION FOR THE TREATMENT OF CHILDHOOD
HEPATIC MALIGNANCIES, TOXICITY, AND EFFICACY: A PILOT STUDY
CHILDRENS CANCER STUDY GROUP, LOS ANGELES, CALIFORNIA

Jorge A. Ortega, Williams Woods, James Feusner,
Gregory Reaman, Beverly Lange, and G. Denman Hammond

Children's Hospital
4650 Sunset Boulevard
Division of Hematology Oncology
Los Angeles, California 90027

INTRODUCTION

The beneficial effect of adjuvant chemotherapy to surgery in children with resectable hepatoblastoma and hepatocellular carcinoma has been documented.[1] However, the same chemotherapy regimens have failed to improve survival for patients with unresectable disease.

Approximately 85% of patients with tumor completely resected at the initial surgery are disease-free two years after diagnosis.[2] However, a response rate of 40% lasting from 3 to 25 months with only a 12% disease-free survival at 2 years has been reported for those with unresectable disease. Despite the demonstrated necessity for complete surgical excision, only 49% of children with hepatoblastoma are able to have their tumor totally removed at diagnosis.[3] In an effort to increase the number of patients with hepatoblastoma and hepatocellular carcinoma receiving the benefits of complete surgical excision, a pilot study was undertaken at a few Childrens Cancer Study Group institutions. For this purpose, repeated courses of adriamycin administered as a continuous I.V. infusion either singly or in combination with cis-platinum was selected.

MATERIALS AND METHODS

The patient population consisted of a total of eleven children with primary hepatic malignancies: six children had hepatoblastoma, all six were under two years of age at diagnosis. Five patients with hepatocellular carcinoma were entered to the study. Of the eleven patients, four had previously received adriamycin as an I.V. bolus.

All patients received the adriamycin as an I.V. continuous infusion for 96 hours. Five patients, three with hepatoblastoma and two with hepatocellular carcinoma received adriamycin infusion in combination with cis-platinum. The six hepatoblastoma patients received a total of 27 courses. Five patients with hepatocellular carcinoma received a total of 20 courses of adriamycin I.V. continuous infusion.

Of the 47 courses of adriamycin I.V. continuous infusion administered, 20 were in combination with cis-platinum at the dose of 100 mg/m^2 over 4 hours infusion. The total adriamycin dose received by this group of patients as I.V. infusion ranged from 180 to 720 mg/m^2 with a median of 360 mg/m^2 and a mean of 385 mg/m^2.

Prior to each adriamycin I.V. infusion course, the following investigations were obtained: CBC, Platelet count, Urinalysis, Total bilirubin, Alkaline phosphatase, SGOT, SGPT, PT, PTT, Serum creatinine, BUN, Electrolytes, Ca, Mg, Creatinine clearance. Echocardiogram and/or radionuclide angiography. Disease evaluation by CT scan and/or ultrasound was performed every other chemotherapy course.

Table I summarizes the patient's characteristics, the adriamycin dose they received and their responses to therapy.

RESULTS

Of the six hepatoblastoma patients, four are presently alive and free of active disease and of these four, two had surgical resection of their tumor after two courses of adriamycin. One patient obtained a good partial response after four courses of the chemotherapy and one patient died of progressive disease after one course of the drug. Of the five patients with hepatocellular carcinoma treated, one patient is presently without demonstrable disease after six courses of adriamycin infusion with cis-platinum. Three good responses lasting for 6, 8 and 10+ months were reported. One patient died of progressive disease after one course of adriamycin-cisplatinum.

TABLE I

PT	CLINICAL CHARACTERISTICS	ADRIAMYCIN DOSE/# OF COURSES/TOTAL DOSE	RESPONSE
HTB			
1	Recurrent with pulmonary mets.	22.5 mg/m^2/day x 4 days/9/647 mg	CR
2*	Recurrent	22.5 mg/m^2/day x 4 days/3/350 mg	CR
3*	Recurrent with pulmonary mets.	22.5 mg/m^2/day x 4 days/1/180 mg	PD
4	Newly diagnosed	22.5 mg/m^2/day x 4 days/8/720 mg	CR
5	Unresectable with pulmonary mets.	22.5 mg/m^2/day x 4 days/4/360 mg	GPR
6	New diagnosed, unresectable	22.5 mg/m^2/day x 4 days/2/180 mg	CR
H-C Ca			
1	Partial resection	22.5 mg/m^2/day x 4 days/2/180 mg	PR
2	New	16 mg/m^2/day x 4 days/7/504 mg	PR
3	Unresectable	22.5 mg/m^2/day x 4 days/4/360 mg	PR
4*	Recurrent	15 mg/m^2/day x 4 days/1/310 mg	PR
5	Partially resected	20 mg/m^2/day x 4 days/6/440 mg	CR

*Previous adriamycin as I.V. bolus.

CR=complete remission; GPR=good partial remission; PD=progressive disease

TOXICITY

Severe neutropenia with absolute neutrophil count below 500/mm^3 was the most common undesirable effect. It was observed in ten different occasions:[2] nine of which were following courses given at the dose of 22.5 mg/m^2/day and five in combination with cis-platinum. One episode of documented sepsis was reported. Severe mucositis complicated by diarrhea was reported in three occasions. Two of these episodes were associated with significant electrolyte imbalance, one requiring parenteral nutrition for three weeks. Moderate vomiting was reported in two instances. Cardiac toxicity was not observed and minimal liver toxicity consistent with mild SGOT elevation was reported in one instance.

DISCUSSION

At the present time there is no widely accepted, standard therapy for the hepatic malignancies of childhood. Total surgical resection even when[3] potentially curative is only feasible in a minority of patients. Systemic chemotherapy can induce responses, but these responses are usually not long lasting. Of all chemotherapeutic drugs adriamycin as a single agent appears to be the most effective for pediatric hepatic malignancies. Unfortunately, there are few reports of the use[4] of adriamycin as a single agent in pediatric hepatoblastomas.

The Childrens Hospital of Los Angeles[5] recently reported their experience with the use of preoperative chemotherapy in an attempt to decrease the size of the tumor in children with primary hepatic malignancies. Seven of the eight children treated exhibited a pronounced clinical response and, four were able to have complete uncomplicated surgical excision of residual disease. Of these, three are alive and without incidence of disease after completion of all therapy. All patients in that study received adriamycin and one patient as a single agent; the rest of the patients received it in combination with other chemotherapeutic agents. St. Jude Children's Research Hospital has also reported[6] on the efficacy of cis-platinum in patients with unresectable or recurrent hepatoblastoma. Most recently, Quinn at the University of Connecticut Health Center[7] reported on the beneficial results obtained with two patients with nonmetastatic unresectable hepatoblastoma treated with adriamycin and cis-platinum.

The finding by Ritch et al.[8] that duration of drug exposure is a major determinant of adriamycin induced tumor cell lethality encouraged us to use the agent as a continuous I.V. infusion for the treatment of childhood liver malignancies. Furthermore, the antracycline induced cardiomyopathy a major limitation to the continued use of the drug beyond a total dose of 450 mg/m^2, has been shown[9] to be significantly reduced with the 96 hours infusion. Results of this preliminary trial demonstrates that adriamycin given as an I.V. continuous infusion is an effective therapy for pediatric liver tumors and that it can be administered safely to children with liver malignancy. Adriamycin continuous I.V. infusion in combination with cis-platinum may constitute the most effective chemotherapy for childhood hepatic malignancies.

REFERENCES

1. A. E. Evans, V. S. Land, W. A. Newton, J. G. Randolph, H. N. Sather, and M. Tefft, Combination chemotherapy (vincristine, adriamycin, cyclophosphamide and 5-fluorouracil) in the treatment of children with malignant hepatoma. Cancer 50:821-826 (1982).
2. A. Ablin, Summary Study Report CCG-881: Hepatoblastoma-hepatocellular carcinoma in children. February 1983, CCG Report. Privileged communication.
3. P. R. Exelby, R. M. Miller, J. L. Grosfeld, Liver tumors in children in the particular reference to hepatoblastoma and hepatocellular carcinoma. J. Pediatr. Surg. 10:329-337 (1975).
4. C. Tan, Adriamycin in pediatric malignancies. Cancer Chemotherapy Reports 6:259-266 (1975).
5. M. E. Weinblatt, S. E. Siegel, M. M. Siegel, P. Stanley, and J. J. Weizman, Preoperative chemotherapy for unresectable primary hepatic malignancies in children. Cancer 50:1061-1064 (1982).
6. J. Champion, A. A. Green, C. B. Pratt, Cisplatin (DDP) an effective therapy for unresectable or recurrent hepatoblastoma. ASCO 671:173 (1982).
7. J. Quinn, A. Altman, L. Garcia, T. Robinson, S. Foster, R. Cooke, D. Hight, Adriamycin (Adria) and cis-platinum (DDP) for unresectable hepatoblastoma. ASCO 299:77 (1984).
8. P. S. Ritch, S. J. Occhipinti, K. S. Skramstad, and S. E. Shackney, Schedule optimization of adriamycin in sarcoma 180 in vitro. Proc. AACR 20:61 (1979).
9. S. S. Legha, R. S. Benjamin, B. Mackay, M. Ewer, S. Wallace, M. Valdivieso, S. L. Rasmussen, G.R. Blunenschein, E. J. Freireich, Reduction of doxorubicin cardiotoxocity by prolonged continuous intravenous infusion. Annals of Internal Med.

AN UNCONTROLLED PHASE II STUDY OF CONSTANT INFUSION

VINCRISTINE-ADRIAMYCIN

L. Helson, M. A. Castello*, E. Arenson**,
L. Steinherz, and S. Groshen

Memorial Sloan-Kettering Cancer Center
New York, U.S.A., *University of Rome, Italy; **Albany
Medical College, New York, U.S.A.

INTRODUCTION

The combination of adriamycin and vincristine administered as bolus injections have antitumoral activity in pediatric cases; however there are no published reports of concomitant constant infusion vincristine and adriamycin. Due to their short serum half-life, it seems reasonable that an improvement in their therapeutic index might occur if these drugs were given as a continuous infusion. In support of this, there is published evidence of reduced cardiotoxic effects in adults for adriamycin when it was administered as a constant effusion (1). We considered it reasonable to evaluate the clinical effects of the combination administered as a constant infusion in patients who were considered refractory to these same drugs given as bolus injections.

We report here the toxicity and responses to the combination vincristine and adriamycin given as a three-day constant infusion in patients with neuroectodermal tumors, Wilms' tumor and hepatoblastoma who have failed to respond to, or who relapsed after, prior treatment with drug combinations including bolus vincristine and adriamycin.

MATERIALS AND METHODS

From September 1984 to April 1985, fourteen patients or their parents were approached and gave signed informed consent in accordance with Federal guidelines to participate in this constant infusion vincristine-adriamycin protocol study. The protocol was approved by the Memorial Hospital Institutional Review Board.

Patients were required to have histologic proof of cancer and previous treatment with vincristine "and/or" adriamycin given as bolus doses in a pulsed schedule to be eligible for study. Other entry requirements included measurable tumor(s) or biochemical products of tumors, such as alpha feto-protein and catecholamines; life expectancy of six weeks; adequate cardiac, hepatic, renal functions; and previously administered adriamycin below a cumulative 460 mg/m^2. Patients were not to have received any chemotherapy and/or radiotherapy for two previous weeks and must have recovered from reversible effects of prior chemotherapy.

The initial evaluation included physical examination, estimation of Karnofsky performance status, EKG and echocardiogram, chest radiograph, blood chemistry, hematologic profile, radionuclide scans as clinically indicated and urine catecholamines or serum alpha-feto protein. A course consisted of vincristine, 1 mg/M^2/day and adriamycin, 15 mg/M^2/day perfused for 24 hours daily through a central line for three consecutive days. The upper limit of vincristine given to any patient was 1 mg/day for three days. Patients were treated with a minimum of two courses. Further courses were administered in the presence of response or stopped when progression of disease or cumulative toxicity was established. After fourteen patients had been assessed and completed two or more courses, we considered our initial target sample to be achieved. Based upon this, the decision to report this study was made. Interim analyses of patients' responses and toxicity following each treatment were recorded independently, in order to determine whether additional courses should be given.

Fourteen patients with histologically proven neuroblastoma (6), phaeochromocytoma (1), esthesioneuroblastoma (2), primitive neuroectodermal tumor (PNET) (2), medulloblastoma (1), hepatoblastoma (1) and Wilms' tumor (1), were entered on study. Of the fourteen patients ranging in age from two to 28 years of age entered and eligible for this study, all had pathological confirmation of cancer and with the exception of one patient who had not received previous adriamycin, all were considered refractory to vincristine and adriamycin given as bolus doses.

MONITORING DURING STUDY

An interval history, physical examination, echocardiogram, EKG and biochemical analyses were obtained prior to each course of treatment. Complete blood count and platelets were monitored twice weekly between courses. Tumor response studies were performed prior to each course or as indicated by physical examination. The Memorial Hospital toxicity rating scale was used. Grades 2 or greater were considered to be clinically significant. Memorial Hospital standard criteria of response were utilized in this study to evaluate the antitumor effects of the drugs. The data generated in this study included objective signs of response, detected by surgery, pathological and biochemical analyses, radionuclide scanning and radiographic examination, and objective signs of toxicity demonstrated by serologic, hematologic, electrocardiographic and sonographic examination. Subjective signs and symptoms were elicited and recorded by MC and EN for patients #1 and #12, respectively, and LH for the remainder. All patients were presented and eligibility discussed prior to entry, at the Pediatric Departmental Tumor Board. Records of this admission to the trial were entered in the patient's chart and the Pediatric pharmacy where the drugs were formulated and placed into mechanical devices for constant infusion. The data were also entered into a separate set of flow sheets which complemented the above systems. Independent analyses of echocardiographic and electrocardiographic changes were done by LS.

Patients entered on study were considered eligible for analysis of response only if they completed two courses and survived an additional two weeks. Patients completing only one course and surviving for two weeks were considered eligible for analysis of toxicity. Of the fourteen patients entered, all fulfilled response and toxicity criteria requirements.

RESULTS

Non-hematological Toxicity

Fifty-five courses of constant infusion vincristine-adriamycin administered to fourteen patients were evaluable for non-hematological toxicity. Nausea and vomiting were seen in two patients for six courses and two courses, respectively; stomatitis was observed in two patients in four of eight courses. No drug related constipation or diarrhea was noted. Alopecia developed in all patients who were not already affected within two weeks of the first course. There were no hepatic, renal, pulmonary, or neurologic toxicities beyond augmented decrease in deep tendon reflexes attributable to the vincristine. Exacerbation of a mild chronic inappropiate anti-diuretic hormone syndrome secondary to carbamazepine was attributed to the first vincristine infusion in one patient with esthesioneuroblastoma, however, a second course of vincristine and adriamycin without carbamazepine was not accompanied by these side effects. Although some of these patients had abnormal echocardiographic findings at entry to study, such as mildly reduced ejection fractions, there were no significant or minimal pericardial effusion. There was no exacerbation or worsening of these following treatment with vincristine and adriamycin.

Hematological Toxicity

The extent of and specific type of myelosuppression differed among the patients. Severe grades of pancytopenia or anemia, leukopenia, or thrombocytopenia were more common in patients who had extensive previous treatment. Variability in myelosuppression in individual patients in different courses was also observed. In general, the degree of leukopenia was less severe than anemia and thrombocytopenia. Although two positive septicemic episodes occurred, no deaths in this patient population could be attributed to infection secondary to leukopenia. The nadir of the leucocyte count was reached within a mean of eight days and the mean time to return to normal levels was the fifteenth day.

Antitumor Activity

One patient with metastatic Wilms' tumor and one with PNET obtained a complete remission (CR). The duration of CR (normal physical examination and radiographic studies and normal markers for over one month) was 7+ months for the Wilms' tumor patient. The patient with the PNET tumor has a CR documented by resection of a residual necrotic tumor after four courses of therapy. She remains free of disease for 4+ months. Two neuroblastoma, one esthesioneuroblastoma, one hepatoblastoma, and one medulloblastoma patients had partial responses (PR) (over 50% decrease in size of tumor masses and/or over 90% decrease in previously increased markers lasting over one month). The remaining seven patients had stable or progressive disease.

These data suggest that vincristine and adriamycin in moderate doses given as a constant infusion in both children and adults is well tolerated. The accompanying side effects are mainly limited to hematologic toxicity and not beyond that usually observed with pulsed injections of adriamycin. The fifty percent objective and partial tumor regressions in this study suggest that continuous infusion vincristine and adriamycin given at the doses and schedules used in this trial may have equal or possibly enhanced antitumor activity compared to similar doses given as pulsed injections. Since the tumors in these patients were considered to be refractory to pulsed doses of vincrisitne and adriamycin, this implies that the clinical effects of this drug combination might be more effective in patients who were not previously treated with these or similar drugs.

The mechanism responsible for the enhanced antitumor effect seen in these patients may be due to the additive effect of both drugs being available to tumor cells during the limited susceptible phases of their cell cycle. Alternatively, or in addition, it is possible that both drugs share or compete for energy dependent efflux mechanisms; hence, for a given dose, greater amounts of both may be retained within the tumor cells. Similar conjecture has been alluded to the role of calcium channel blockers in augmenting adriamycin and vincristine tumoricidal activity in p388 cells resistant to adriamycin and vincristine (2). Enhancement of cardiac cell accumulation of adriamycin when given simultaneously with vincristine is unknown but may be clinically insignificant based upon the absence of enhanced cardiotoxic effects seen in these patients. The patient with metastatic medulloblastoma also had symptomatic co-existant sub-aortic hypertrophic stenosis. Verapamil was given for therapeutic reasons independent of its adriamycin-vincristine enhancing effects, and on serial echocardiographic and electrocardiographic studies, no deleterious effects were noted. This patient also exhibited a good partial response to treatment (Figs 1,2). Nevertheless, the potential for enhanced cardiotoxicity in patients given vincristine, adriamycin and verapamil concomitantly cannot be entirely excluded in the absence of application of endomyocardial biopsies, or more sensitive indicators than echocardiograms which for practical purposes were omitted from this trial.

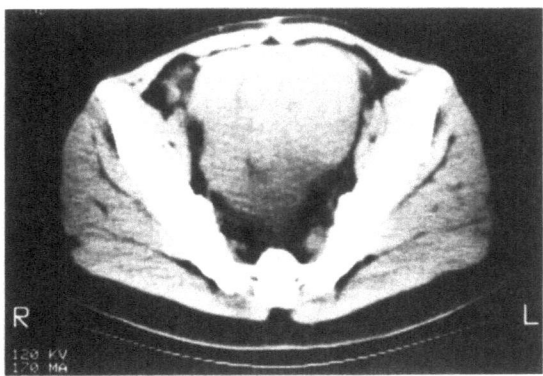

Fig. 1: Computerized tomography of the pelvis of a 28-year-old male with disseminated medulloblastoma. The tumor extended beyond the pelvis into the lower abdomen. This tomograph was taken following conventional chemotherapy which included pulse doses of vincristine-adriamycin; and was at a time during which his tumor was increasing.

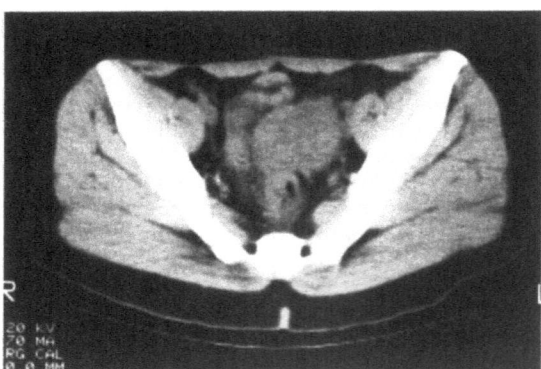

Fig. 2: Computerized tomography of the pelvis of the patient in Fig. 1; after four courses of constant infusion vincristine-adiramycin. The tumor has diminished in size to less than 50% of its original size.

BIBLIOGRAPHY:

1. Legha S.S., Benjamin R.S., Mackay B, et al: Reduction of doxorubicin cardiotoxicity by prolonged continuous intravenous infusion. Ann Int Med 96:133-139, 1982.

2. Helson L: Calcium channel blocker enhancement of anticancer drug cytotoxicity - A review. Cancer Drug Delivery 1(4):353-361, 1984.

LOW-DOSE ARA-C BY CONTINUOUS INFUSION IN THE
TREATMENT OF ACUTE NON-LYMPHOCYTIC LEUKEMIA
(ANLL) AND MYELODYSPLASTIC SYNDROME (MDS)

Farida Chaudhri, Steven L. Allen, Philip Schulman,
Willi Kreis, Daniel R. Budman, Lora Weiselberg,
and Vincent Vinciguerra

Don Monti Division of Oncology, Department of Medicine
North Shore University Hospital, Manhasset, N.Y. 11030
Cornell University Medical College, N.Y., N.Y. 10021

ABSTRACT

Thirteen patients with acute non-lymphocytic leukemia (ANLL) or myelodysplastic syndrome (MDS) were treated with low-dose cytosine arabinoside (Ara-C) 20 mg/m^2/day by continuous intravenous infusion for 21 days. Seven patients attained complete remissions (CR) and two patients had partial remissions (PR). Two patients required two courses of treatment before attaining CR. Two of four patients with abnormal cytogenetics also achieved complete response with reversion of cytogenetic abnormalities to normal. All patients who responded showed profound pancytopenia with marked hypoplasia of bone marrow. The universal occurrence of pancytopenia and reversion of cytogenetic abnormalities to normal suggests that low-dose Ara-C is cytotoxic and may not act as a differentiating agent.

INTRODUCTION

Cytosine arabinoside is the mainstay of modern treatment for ANLL. However, in cases where this treatment is contraindicated or ineffective, novel approaches are required. Drug dosage or schedule can be manipulated to exploit the drug's pharmacokinetics. Ara-C is an S-phase specific agent. It should therefore be possible to design a schedule which maintains an effective serum concentration of Ara-C for the time period required for all leukemic cells to progress into the sensitive S-phase.

Sublethal doses of Ara-C have been shown to induce terminal differentiation in vitro of human myeloid leukemic cells (1-3). Therefore, it is possible that low doses of this drug might induce differentiation in vivo of leukemic cells to mature forms without marked toxicity, avoiding the risks of pancytopenia associated with conventional dosages of Ara-C.

These theoretical advantages might be especially important in several common clinical situations. Elderly patients with ANLL frequently do not tolerate the complications of standard therapy, particularly those associated with severe, prolonged pancytopenia. Patients with MDS also cannot tolerate conventional chemotherapy and generally do not benefit from it (4, 5, 6).

Recent evidence suggests efficacy for low dose Ara-C in patients with MDS and ANLL (7-11). Encouraged by these results and to clarify the mechanism of action of this therapy, a phase II study of low-dose Ara-C delivered by continuous intravenous infusion was employed in 13 patients with ANLL and MDS.

METHODS

Adult patients with ANLL in whom standard aggressive therapy was relatively contraindicated due to advanced age or had been ineffective in obtaining complete remission were included in this study. In addition, patients with MDS were also treated with this regimen. Cytogenetic analysis was performed on all patients prior to treatment and upon achievement of CR. Specific criteria for inclusion in the study included: 1) age 16, 2) performance status >40%, 3) documented myelodysplasia or ANLL. Those with MDS were required to have a significant cytopenia and anemia refractory to standard therapy for at least six months. The patients were subclassified according to criteria set forth by the French-American-British Cooperative Group for patients with ANLL and MDS (12).

Informed consent was obtained from all patients. This study was approved by the Institutional Review Board of North Shore University Hospital.

All patients were hospitalized and had insertion of a triple lumen catheter in a subclavian vein prior to treatment. Ara-C 20 mg/m^2/day was given by continuous intravenous infusion using an IVAC pump for 21 days. Patients who did not respond to the initial course of therapy received a second course two weeks after the first. If complete remission was obtained, no maintenance therapy was given.

Complete response (CR) was defined as normal peripheral blood counts (leukocytes, hemoglobin, platelets, differential count) and peripheral smear with the bone marrow demonstrating < 5% blasts with normal cellularity and cytogenetics. Partial response (PR) was defined as an improvement in peripheral blood count with leukocytes > 1.5 x 10^9/l, and hemoglobin $\geq$ 10 g/dl with no further transfusion requirements and the bone marrow containing > 5% blasts.

RESULTS

Thirteen patients with ANLL or MDS were treated between November 1983 and January 1985. Patient characteristics are as shown in Table I.

Table I PATIENT CHARACTERISTICS

Age: Median, 68; Range 26-84
Sex: 7M, 6F
Performance Status: Median 80%; Range 40-90%

Diagnosis	No. of Patients
ANLL (de novo)	5
MDS	4
Secondary ANLL	3
Acute Myelofibrosis	1

Pre-treatment Cytogenetics

  6 NN
  4 AN   -5, 5q-, r7, i (21), 5 (11,20)
  2 AA   +8

Prior Therapy

  Yes - 2
  No - 11

The age range of patients was 26-84 years, with a median age of 68 years. The male:female ratio was 7:6.

Patients were subclassified into four categories: 1) 5 cases of ANLL de novo (2 M1, 1 M2, and 2 M5), 2) 4 cases of MDS, 3) 3 cases of secondary leukemia following chemotherapy (2 Hodgkin's disease and 1 adjuvant chemotherapy for breast cancer), 4) 1 case with acute myelofibrosis.

Two patients had had prior chemotherapy for leukemia. One was treated with conventional doses of cytosine arabinoside and achieved a CR but relapsed 14 months later. The other patient was treated with subcutaneous low dose Ara-C for MDS for 6 months with improvement in blood counts, but subseuqently developed ANLL.

Responses to low dose Ara-C were observed in all categories of patients and are summarized in Table II.

Table II RESPONSE BY DISEASE

Disease	No. Patients	Response
ANLL (de novo)	5	3CR
MDS	4	3CR 1PR
Secondary ANLL	3	1PR
Acute Myelofibrosis	1	1CR

Three out of five patients with ANLL de novo responded. All 4 patients with MDS and 1 of 3 patients with secondary ANLL responded. Three patients were resistant to chemotherapy.

Two patients required two courses of therapy to obtain CR. One was the patient with acute myelofibrosis who achieved a CR after two courses of treatment with complete resolution of fibrosis and karyotypic conversion to normal. This response lasted 7 months prior to relapse.

All patients who responded to treatment had marked reduction (Table III) in peripheral blood counts and marked hypoplasia of the bone marrow.

Table III MARROW RESPONSE

	No. of Pts.	Median Time to Response	
		(days)	(range)
Differentiation	1	14	
Hypoplasia/Aplasia	9	21	(14-35)
Recovery (PMN's > 500) (platelets > 100,000)			

Marrow hypoplasia was generally noted by day 14 of treatment. An increase in maturation with reduction in the number of blasts on consecutive marrow and peripheral blood count recovery occurred approximately two weeks after completion of treatment.

The duration of complete remission has been relatively short (Table IV) with a median of 5 months and a range of 3 to 13 + months.

Table IV

Responses	Duration of Responses (mos)
CR 7	3, 4, 4 ½+, 5, 5½, 7, 13+
PR 2	1, 4+

Six patients remain alive. The median duration of survival in complete responders is 7 months (range 3.5 - 13+ months) compared to 4 weeks in non-responders. Among the two partial responders, one patient died 4 weeks after response due to myocardial infarction and the other patient relapsed after 4 months, but is still alive at 10+ months.

Three out of three patients were successfully retreated with the same low-dose Ara-C regimen following relapse. One died during the recovery phase due to intractable grand-mal seizures. The remaining two patients achieved a second complete remission, with one relapsing again after only 2 months and the other remaining in remission at 4+ months.

Of the 4 patients with abnormal cytogenetics, two achieved a complete remission. Each required two courses of treatment and each had conversion of their karyotypes to normal. One out of the two patients who were previously treated achieved complete response.

TOXICITY

Toxic side effects were significant (Table V).

Table V

TOXICITY (no. patients)

	Grade 0	1	2	3	4
Hematological	-	-	-	-	13
Hepatic	-	3	-	-	-
G. I. Toxicity	-	3	2	-	-
Skin	-	2	-	-	-
Alopecia	-	-	-	-	-
Renal	-	-	-	-	-

Grade 3 to 4 hematologic toxicity occurred in all patients including three treatment-related deaths (2 septic, 1 hemorrhagic). Pancytopenia was noted in all patients after 7-10 days of treatment. Median time from its onset to recovery of peripheral blood counts was 32 days. During this phase, all patients required intensive platelet and packed red blood cell transfusional support. Five patients had documented bacterial infection. Two patients had strong clinical evidence of fungal infection and responded to amphotericin B.

Gastrointestinal toxicity was limited to nausea in three patients and diarrhea in two. Three patients had transient elevation of liver function tests during therapy. Two patients developed a maculopapular rash. Alopecia, nephrotoxicity and mucositis were negligible.

PHARMACOKINETICS

The pharmacokinetics of low-dose cytosine arabinoside were studied in 11 patients. The patients' plasma and urine were analyzed for Ara-C by radioimmunoassay as described by Piall et al (13). The mean plasma concentration of 189 samples collected during steady state (C pss) was 7.7 ± 4.7 ng/ml with a range of 0.6 to 29.7 ng/ml. Mean plasma levels were reached within 2.7 hours and decreased rapidly following discontinuation of the infusion. Other parameters measured included area under the curve (AUC) 182.1 ± 64.8 ng/day/ml, volume of distribution at steady state (VDSS) 53,913 ± 17,626L, total body clearance (TBC) 188.7 ± 54.8L/hr, renal clearance (RC) 3.1 ± 1.4L/hr. The recovery of unchanged Ara-C in the urine was 1.43 ± 0.69% of daily infused Ara-C, the bulk being excreted as Ara-U (measured by high pressure liquid chromatography in selected samples). These values indicate that Ara-C is both extensively distributed and rapidly and almost completely deaminated to Ara-U.

DISCUSSION

Baccarani and Tura (7) described a patient with MDS treated with low-dose Ara-C (0.7 mg/kg/day) by intravenous bolus for seven days. The bone marrow improved and peripheral blood counts became normal. This remission lasted six months. Housset et al (8) described three patients treated with low dose Ara-C, two with ANLL de novo and one with acute transformation of MDS. All three patients achieved complete remissions which lasted more than six months. Castaigne et al (10) reported responses in four of five patients with MDS and 11 of 16 patients with ANLL who were treated with Ara-C 10 mg/m^2/12 hours by subcutaneous injection for 17-25 days. Twelve patients achieved CR and three patients had partial responses.

Wisch et al (11) recently demonstrated efficacy of low dose Ara-C in eight patients with preleukemia or with preleukemia in evolution to acute leukemia. All patients were treated with Ara-C 20 mg/m^2/day by continuous infusion for 7-21 days. Six of eight patients had marked increases in peripheral blood granulocyte, platelet and red cell counts with concomitant reduction in transfusion requirements.

In our study, Ara-C 20 mg/m^2/day by continuous intravenous infusion over 21 days was administered to 13 patients with ANLL or MDS in whom standard chemotherapy was contraindicated or had been ineffective. Seven patients attained a CR and two had a PR, confirming the efficacy of low dose Ara-C by continuous infusion in ANLL and MDS. Of note, by definition, our complete responses required normalization of the hemogram, bone marrow, and conversion of abnormal cytogenetic findings to normal.

The mechanism by which low dose Ara-C induces responses in patients with MDS and ANLL is unclear. Lotem and Sachs (1) demonstrated that murine myeloid leukemia cells cultured in vitro can be induced to differentiate to mature forms after incubation with low doses of various agents, including Ara-C. The authors of subsequent case reports (7-11) of induction of remission of ANLL and MDS by low dose Ara-C tended to favor terminal cell differentiation as the mechanism of action of low dose Ara-C. In contrast, all of our patients had profound pancytopenia and marked marrow hypoplasia. Only one patient showed evidence of a differentiating effect, with consecutive marrows demonstrating an increasing number of mature cells with persistent, but decreasing numbers of myeloblasts. Furthermore, of the two patients with abnormal cytogenetics who achieved a CR, both had karyotypic conversion to normal. These results suggest that low dose Ara-C may actually function as a cytotoxic agent, similar to standard doses of the drug, rather than as a differentiating agent.

Duration of responses in this study was short-lived, with a median of 5 months. However, two patients are still in complete remission after 11+ and 14+ months. Repeated courses of maintenance therapy might allow prolongation of the duration of response. In this regard, it is noteworthy that two of three patients who were re-treated with low dose Ara-C following relapse achieved a second complete remission.

A major drawback of this treatment regimen is the prolonged period of hospitalization required, generally 4-8 weeks. This is similar to the length of stay necessary for standard treatment regimens of ANLL. The use of portable transfusion pumps may allow the period of hospitalization to be reduced. Treatment could be initiated on an out-patient basis with admission to the hospital reserved until infectious or hemorrhagic complications necessitates it.

In summary, our study suggests that the continuous infusion of low dose Ara-C is effective therapy for ANLL and MDS. This approach should be considered primarily in cases where standard treatment is either ineffective or contraindicated. It is well tolerated by the elderly, but should not be considered benign in view of the marked pancytopenia which is induced. The mechanism of action of low dose Ara-C by continuous infusion appears to be mediated via cytotoxicity and not via differentiation.

REFERENCES

1. Lotem, J., Sachs, L. Different Blocks in the Differentiation of Myeloid Leukemic Cells. Proc. Nat. Acad. Sci. 1974, 71:3507:3511.
2. Krystosek, A., Sachs, L. Control of Lysozyme Induction in the Differentiation of Myeloid Leukemic Cells. Cell 1976: 9:675-584.
3. Sachs, L. The Differentiation of Myeloid Leukemia Cells: New Possibilities for Therapy. Br. J. Haem. 1978; 40:509-517.
4. Cohen, J. R., Creger, W. P., Greenberg, P.L., Schrier, S.L. Subacute Myeloid Leukemia: A Clinical Review. Am. J. Med. 1979: 66:959-66.
5. Crosby, W.H. To Treat or Not to Treat: Acute Granulocytic Leukemia. Arch. Int. Med. 1968; 122:79.
6. Cooperative Group for the study of aplastic and refractory anemia with excess of blast cells: Prognostic factors and effect of treatment with androgen or Cytosine Arabinoside: Results of prospective trial in 58 patients. Cancer 1979; 44:1976-82.
7. Baccarani, M., Tura, S. Differentiation of Myeloid Leukemic Cells: New Possibilities for Therapy. Br. J. Haem. 1979; 42:485-487.
8. Housset, M., Daniel, M.T., Degos, L. Small Doses of Ara-C in the Treatment of Acute Myeloid Leukemia: Differentiation of Myeloid Leukemia Cells. Br. J. Haem. 1982; 51:125-129.
9. Maloney, W.C., Rosenthal, D.S. Treatment of Early Acute Non-lymphatic Leukemia with Low Dose Cytosine Arabinoside. Haem. and Blood Trans. 1981; 26:59-62.
10. Castaigne, S., Daniel, M.T., Degos, L. Does Treatment with Ara-C in Low Dosage Cause Differentiation of Leukemic Cells? Blood 1983; 62:85-86.
11. Wisch, J.S., Giffin, J., Kufe, D. Response of pre-leukemic syndromes to continuous infusion of low dose Ara-C. N. England J. Med. 1983; 239:1599-1605.
12. Bennett, J.M., Catovsky, D., Daniel, M.T., Flandrin, G., Galton, D.A.G., Gralnick, H.R., Sultan, C. The French-American-British (FAB) Cooperative Group. Proposals for the Classification of the Myelodysplastic Syndromes. Br. J. Haem. 1982; 51:189-199.
13. Piall, E.M., Aherne, G.W., Marks, V.M. A Radioimmunoassay for Cytosine Arabinoside. Br. J. Cancer 1979; 40:548-556.

THE 5-DAY CONTINUOUS INFUSION OF CIS-PLATINUM: AN UPDATE ON TOXICITY PATTERN

P. Salem, M. Khalyl, K. Jabboury, L. Hashimi

American University of Beirut Medical Center

Beirut - Lebanon

Cis-Diamminodichloroplatinum (II), or DDP, has been shown to be an effective agent in the treament of various neoplastic diseases. The drug has been usually administered in high intravenous (IV) bolus dose, or rapid infusion. By these schedules, the toxicity of DDP included severe nausea and vomiting, nephrotoxicity, mild to moderate myelosuppression, and hearing loss[1]. Renal and gastrointestinal toxicities were dose-limiting and constituted major obstacles to prlonged therapy. Although nephrotoxicity was partially ameliorated by fluid and mannitol diuresis[2], it remains a major impediment of long-term treatment. Also, in several studies, nausea and vomiting were so severe that many patients refused to continue treatment.

In an attempt to further diminish the toxicity of DDP, clinical trials with the drug administered by 5-day continuous IV infusion (CIVI), were initiated in 1976[3]. In regard to the anti-cancer effect of such a dose schedule, Drewinko et al[4]. observed that DDP cytotoxicity to asynchronous human lymphoma cells in culture was enhanced by prolonged exposure to the drug. In addition, Sidgestad et al.[5] had observed that cells are most vulnerable to the lethal effect of DDP while in the G1 phase of the life cycle. Also, DeConti et al.[1] showed that 90% of the platinum administered was protein-bound within 2 hours. Since the drug fraction that exerts the anti-cancer effect is the unbound moiety and the binding is only very slowly reversible,[1] thus the constant infusion should theoretically produce a greater cell kill by exposing more cells in the G1 phase to an adequate level of the free cisplatin. However, to date, there are no randomized clinical trials that confirm the therapeutic superiority of prolonged infusions over rapid infusions.

MATERIALS AND METHODS

Patient Selection

Ninety-six patients with a variety of histologically proven neoplastic diseases were entered in this study. All patients had a serum creatinine < 1.5 mg/dl. Prior treatment was discontinued 3 to 4 weeks before initiation of cisplatin (DDP).

DDP Administration

Patients were admitted to the hospital and given 3 L of intravenous fluids in the 12 hours preceding treatment. DDP was administered in physiologic saline and infused in 20 mg/m^2 as a continuous 24-hour daily infusion for 5 consecutive days. Courses were repeated every 4 to 6 weeks. During the 5-day continuous infusion course, patients were adequately hydrated. Complete blood counts, platelets, blood urea nitrogen, serum creatinine levels, and electrolytes were monitored every other day, and the contents of the IV fluids were adjusted accordingly. No mannitol or diuretics were given. Subsequent to treatment, relevant blood and biochemical studies, in addition to tumor parameters, were monitored at 2-week intervals.

Evaluation of Toxicity

Gastrointestinal toxicity was evaluated daily by a well-trained oncology nurse as follows: Grade 0: none; Grade 1: nausea without vomiting; Grade 2: transient vomiting; Grade 3: vomiting requiring therapy; and Grade 4: intractable vomiting. Nephrotoxicity was defined as elevation in serum creatinine to more than 2 mg/dl. Serum magnesium was determined by the atomic absorption method and hypomagnesemia was defined as a serum magnesium ≤ 1 mg/dl.

RESULTS

Ninety-six patients received DDP, alone (20 patients) or in combination with other anti-cancer agents.

Toxicity (Table 1)

Gastrointestinal. The only gastrointestinal toxicity encountered in our study was nausea and vomiting. Seven percent of the evaluable cycles were associated with Grade 3 toxicity, while in 78%, this toxicity was Grade 0-1 (Table 2). Twenty-nine percent of the evaluable patients had no nausea or vomiting (Table 1). There was no accentuation of toxicity with subsequent courses or within courses over the 5-day treatment.

Table 1. General Pattern of Toxicity Produced by 5-Day DDP Infusion

Toxicity	% Evaluable patients
None	20
Nausea and vomiting	71
Myelosuppression	69
Hypomagnesemia	30
Nephrotoxicity	5
Ototoxicity	2

Renal. Four (5%) of 81 evaluable patients developed nephrotoxicity (Table 1). There was no apparent correlation between nephrotoxicity and the number of courses received by an individual patient. Indeed, none of the patients who received more than 5 courses developed such toxicity. In all patients who developed nephrotoxicity, there were factors other than DDP which might have contributed, at least partially, to the renal damage (Table 3). One patient (MH), who was recently reported, developed fatal nephrotoxicity as a probable complication of DDP and cephalothin-gentamicin therapy.[6] In the remaining three patients, the highest recorded serum creatinine was 3.3 mg/dl; and in all, creatinine returned to normal in a few months.

Table 2. Gastrointestinal Toxicity Produced by 5-day DDP Infusion

	Cycles	%
Total	280	
Evaluable	256	
Grade		
0	107	42%
1	93	36%
2	38	15%
3	18	7%
4	0	0

Hematologic. Severe anemia (hemoglobin < 8 g/dl) occurred in 23% of 74 evaluable patients, while 53% had hemoglobin < 10 g/dl. The relationship of anemia to DDP therapy, primary disease, or treatment other than DDP could not be clearly delineated. None of the 20 patients who received DDP alone developed neutropenia-related infections.

Hypomagnesemia. Twenty-one percent of 140 evaluable cycles were associated with a serum magnesium $\leq$ 1.3 mg/dl. Serum magnesium dropped to levels $\leq$ 1 mg/dl in only 6 evaluable cycles. This toxicity occurred within the first week of therapy and was always reversible.

Miscellaneous. A small number of patients developed other side effects. One patient developed transient tinnitus, and another had 30% decrease in hearing acuity. One patient developed hallucinations and mental disorientation on day 4 of the first cycle. These symptoms lasted 5 days but their relationship to DDP could not be confirmed.

Table 3. Outcome of Nephrotoxicity and Contributing Factors Other Than DDP in 4 Patients

Creat. peak level (mg/dl)	Date peak level	Probable contributing factors other than DDP	Outcome
7.4	Day 23 Cycle 1	Cephalothin + Gentamicin + Carbenicillin. Sepsis.	Death on day 24
3.3	Day 63 Cycle 5	Severe hypertension + Metastatic renal disease	Recovery 4½ mos.
2.4	Day 34 Cycle 5	Clindamycin	Recovery 8 mos
2.4	Day 23 Cycle 3	Cephalothin + Carbenicillin	Recovery 2½ mos.

DISCUSSION

The objective of our work was to study the toxicity pattern of the 5-day continuous infusion (CIVI) of DDP. Although laboratory studies suggest that prolonged infusions might therapeutically be superior to rapid infusion regimens,[1,4,5] the therapeutic efficacy of CIVI could not be adequately evaluated in this study.

Gastrointestinal toxicity, which poses a major obstacle to prolonged use of DDP, and which occurs in the majority of patients who receive the bolus dose, has been sharply reduced by CIVI. This toxicity was totally lacking in 42% of evaluable cycles, and was severe (Grade 3) in only 7%. This marked decrease in nausea and vomiting is of major clinical usefulness, particularly in patients who refuse to receive further bolus dose treatment because of such toxicity, and also in patients where vomiting is extremely undesirable as those with head and neck or esophageal cancer.

This study also suggests that nephrotoxicity is significantly reduced by CIVI. This toxicity occurred in four patients (5%), and was severe in one, but was mild and reversible in three. In all, there were factors other than DDP, including the concomitant use of nephrotoxic antibiotics, which could have contributed to renal damage. The severe nephrotoxicity which was noted in one patient and which had been earlier described in details,[6] was probably the product of additive renal damage by the administration of cephalothin and gentamicin during Platinum course. This additive toxicity has been previously reported by Gonzalez-Vitale et al.,[7] and therefore, we believe that the use of nephrotoxic antibiotics in the management of patients receiving Platinum should be discouraged. Since this dose schedule is associated only with minimal nephrotoxicity, treatment for more prolonged periods becomes possible. This is particularly important in patients who do not achieve a CR after a few cycles, but who nevertheless continue to show anti-tumor response. Although the majority of patients in our study received 3 cycles or less of cisplatin, 28 patients received 4 or more cycles and only 1 had transient elevation in serum creatinine.

Renal magnesium wasting due to cisplatin has been previously reported.[8] In our study, serum magnesium < 1 mg/dl was encountered in 4% of evaluable cycles. None of our patients developed peripheral neurotoxicity or central nervous system toxicity which was clearly related to cisplatin.

Hematologic toxicity of DDP by CIVI was adequately assessed by one of the authors (P.S.) in a previous study.[3] It was clearly shown that in comparison to the bolus dose, the nadir of the leukocyte count in the CIVI was relatively delayed, and that the median time to recovery was 31 days. Thrombocytopenia and/or granulocytopenia occurred in 69% of evaluable courses.

In conclusion, this new dose-schedule is apparently much less toxic than the bolus IV injection, in terms of immediate and delayed toxicities, and thus allows the use of DDP for more prolonged periods of time. The therapeutic efficacy of this schedule should be compared to the conventional dose-schedules in prospective randomized trials.

REFERENCES

1. DeConti RC, Toftness BR, Lange RC, Creasey WA. Clinical and pharmacological studies with cis-dichlorodiammineplastinum (II). Cancer Res 1973; 33:1310-1315.
2. Hayes D, Cvitkovic E, Golbey R, Schneider E, Krakoff IH. Amelioration of renal toxicity of high-dose cis-platinum-diamminedichloride (CPDD) by mannitol-induced diuresis. (Abstr) Proc Am Assoc Cancer Res 1976; 17:169.
3. Salem PA, Hall SW, Benjamin RS, Murphy WK, Wharton JT, Bodey GP. Clinical phase I-II study of cis-dichlorodiammineplatinum (II) given by continuous IV infusion. Cancer Treat Rep 1978; 62:1553-1555.
4. Drewinko B, Brown BW, Gottlieb JA. The effect of cis-diamminedichloro-

platinum (II) on cultured human lymphoma cells and its therapeutic implications. Cancer Res 1973; 33:3091-3095.
5. Sidgestad CP, Grdina DJ, Peters LJ, Stutesman J. Cell cycle phase preferential killing of fibrosarcoma tumor cells by cis-dichlorodiammineplatinum or Adriamycin (Abstr). Proc Am Assoc Cancer Res 1979; 20:178.
6. Salem PA, Jabboury KW, Khalil MF. Severe nephrotoxicity: A probable complication of cis-dichlorodiammineplatinum (II) and cephalothin-gentamycin therapy. Oncology 1982; 39:31-32.
7. Gonzalez-Vitale JC, Hayes DM, Cvitkovic E, Sternberg SS. Acute renal failure after cis-dichlorodiammineplatinum (II) and gentamycin-cephalothin therapies. Cancer Treat Rep 1978; 62:693-698.
8. Schilsky RL, Anderson T. Hypomagnesemia and renal magnesium wasting in patients receiving cis-platin. Ann Intern Med 1979; 90:929-931.

LONG TERM APPRAISAL OF THE LEMON-FOLEY METHOD OF CHEMOTHERAPY OF SOLID TUMORS OF THE GASTROINTESTINAL TRACT

>Donald Steinberg
>Clinical Associate Professor of Surgery
>Medical College of Ohio at Toledo
>2200 Madison Avenue, Toledo, Ohio - 43624

INTRODUCTION

One of the most difficult problems in the clinical judgement of cancer patients is the determination of the effectiveness of any treatment modality. This is particularly apparent in the evaluation of solid malignancies of the gastrointestinal tract. In 1966, Drs. Lemon and Foley proposed the slow infusion of one gram of 5FU for periods of eight hours daily for 14 days followed by weekly bolus I-V injections of .5 grams 5FU in 10cc of dilutent. They felt that this method was not only effective but also avoided the frequent severe iatrogenic complications of current anti-neoplastic chemotherapy.

For the past 17 years we have followed their recommendations with only slight modification. Our results have also demonstrated a prolonged period of survival in many of these patients with almost no iatrogenic complications of the treatment.

Criteria of Patient Selection

Patients selected for the study were those patients who had either a poor prognosis due to the primary site of the malignancy or were those patients whose extent of disease portended a poor prognosis. Patients of the first group were those with carcinoma of the stomach, anal canal, pancreas and gall bladder. Those of the second type were those with extensions of carcinoma of the colon.

Our criteria of evaluation were the length of independent social existance and also the development of distal metastasis after initiation of 5FU therapy. The absence of distal metastasis after initiation of therapy appeared consistently in the study if they were not present prior to the initiation of 5FU therapy.

Method

Patients selected for the study were those with a poor long term prognosis. Each patient was treated with one gram of 5FU mixed in 1000cc of 5% glucose in water at a rate of 75cc per hour I-V daily for 14 days. Daily white blood count were obtained. If the W.B.C. was gradually falling for three days, the treatment was delayed until the W.B.C. began to rise or the treatment was discontinued if the patient complained of diarrhea or ulceration of the lips or mouth. After completion of the 14 day regime, each patient received .5 gm. 5FU bolus injection on a weekly basis without cessation. No toxicity with this part of the treatment was encountered. These treatments continued for years, the longest being 14 years of almost continuous treatment. One patient has received four courses of 5FU in addition to his weekly injections over 19 months.

Discussion

The number of patients in each primary site is small but as a group they represent a significant improvement in the duration of the periods of palliation. We feel it is also significant that there was a paucity of new areas of metastasis occurring after 5FU therapy was initiated. It would appear that the treatment regimen was effective against that subset of neoplastic cells which were most aggressive. Thus we would see the continuing local extension of carcinoma in some patients without the development of distal metastases. This type of failure to obtain local palliation was best demonstrated in patients with colo-rectal carcinoma. In two patients the use of irradiation to control local recurrence has be a significant factor in continuing the length of palliation. Although it is impossible to document, it seemed that some patients remained at a clinical "status quo" for months despite the presence of massive liver metastasis.

CHART 1

PRIMARY SITE	AGE	SEX	REASON FOR POOR PROGNOSIS	SURVIVAL AFTER D_x	DURATION 5FU	REMARKS
SIGMOID	72	F	Int. Obstrc-Liver Metastasis	1yr 11 mo.	5 months	Severe AutoAccid. R_x disc.
SIGMOID	50	M	Liver Metastasis	7 months	5 months	Good Clinical Pall. 3-4 Months
TRANS. COLON	62	F	Massive Tumor with Fistula into small intestine	18 months	8 months	Progressive deterioration
*RECTUM	67	M	Liver Metastasis	25 months	24 months	Excellent Palliation
*RECTUM	38	M	Liver Metastasis	6 years	9 months	Good Palliation for nine months
RECTUM	64	F	Lung Metastasis	2 years	7 months	Progressive deterioration
RECTO-SIGMOID	60	M	Liver Metastasis	14 months	12 months	Progressive deterioration
RECTUM	58	F	Extension in vagina	4 years	4 months	No distal metastasis but progressive extension into Pelvis
*RECTO-SIGMOID	53	M	Tumor exposed into Peritoneal Cavity	8½ years	7½ years	Alive and well
*RECTO-SIGMOID	69	M	Solitary Liver Metastasis	16 years	14½ years	Alive and well
*STOMACH	46	F	Linitis Plastica	26 months	24 months	Excellent Palliation
*STOMACH	63	M	AdenoCarcinoma with Metastasis to regional nodes	18 months	18 months	Alive and well case present
GALL BLADER	65	M	Massive extension into Liver	8 months	7 months	Excl. Palliation until Biliary Obstruction
*PANCREAS	60	M	Inoperable	1 year	10 months	Malignant effusion after discontinuing treatment
*ANAL CANAL	63	F	Perineal recurrence 2 years Post Bilateral groin dissection with abdominal perineal	5 years	3 years	Carcinomatosis as terminal event
*CARCINOMA IN FISTULA IN ANO	61	F	Massive Pelvic Extension	?	1 year	Doing well - patient stopped treatment - lost to followup

Case Presentation

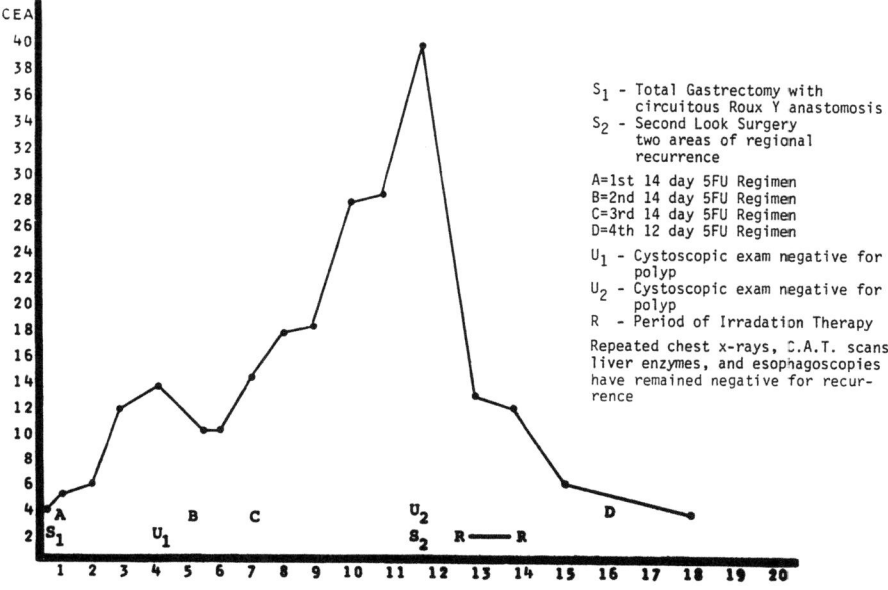

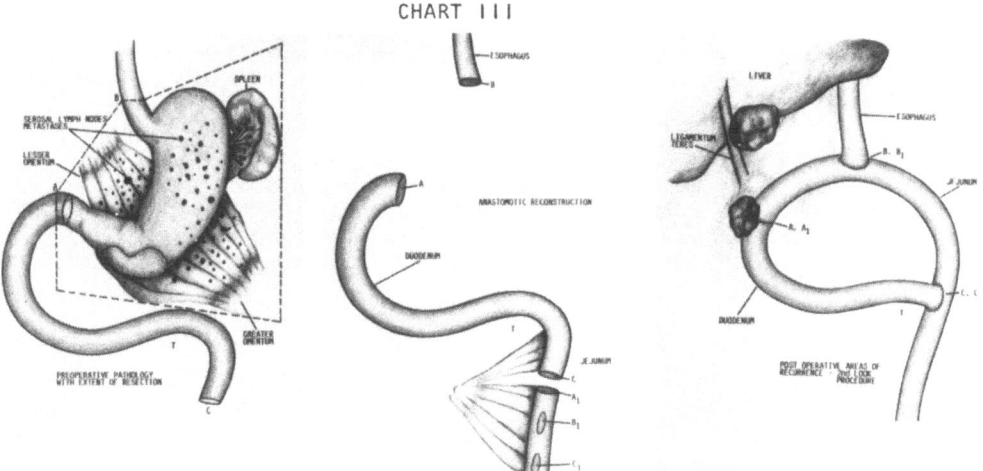

The patient is a 61 year old male admitted to the hospital with upper abdominal pain, increasing anorexia and a 30 lb. weight loss. Upper gastrointestinal examination revealed a large ulceration of the posterior wall and greater curvature of the antrum of the stomach. Gastroscopic exam revealed a large malignant ulcer proven by biopsy. Exploratory laparotomy revealed regional lymph node metastasis as well as nodular metastasis over the serosa of the body and fundus of the stomach. The liver was free of metastasis.

A total gastrectomy with en-bloc excision of the spleen and greater and lesser omentums was performed. Re-establishment of intestinal continuity was accomplished by a circuitous Roux en Y anastamosis. Postoperatively the patient completed a 14 day regimen of 5FU using a modification of the Lemon-Foley method of peripheral I-V infusion for 12-14 hours per day. After discharge the patient received weekly I-V injections of .5 gms 5FU for five months. Repeat CAT Scan, chest x-ray, liver enzyme, etc., were negative for metastasis. Cystoscopic exam revealed that there was no recurrence of bladder polyps that had been removed prior to his presenting illness. Patient was then given a second course of 14 days of one gram of 5FU for 14 days without complication other than a recurrent acute cystitis which was treated by antibiotics. Two months later the patient received his third course of 5FU for 14 days after receiving his weekly 5FU.

Throughout the post-op course the patient had monthly CEA antigen determinations. There was a progressive slow increase in the levels throughout the first post-op year. The CEA rose from 4.2 to 40 at which time a small area of tumefaction was palpable in the upper incisional area. Second look surgery was performed and two areas of recurrence were found. One involved the duodenojejunostomy and the other the ligementum teres extending into the quadrate lobe of the liver. Both areas were excised and also marked with silver clips for post-op radiation localization. Post operative radiation was administered over the designated areas as well as the periaortic areas about the celiac axis. Following surgery and radiation there has been a dramatic fall in the CEA level to 4.8. At the present time there is no evidence of distal metastasis. Repeat cystoscopic exam is still normal with no evidence of polyp formation. At sixteen post-op months the patient received his fourth course of 12 gms. of 5FU in 12 days. The treatment was terminated because of the development of diarrhea.

The patient is now 18 months since his original surgery and is asymptomatic. Weekly I-V injections of 5FU are continuing.

Conclusion

From this preliminary long term study, it appears that 5FU may be an effective therapy in gastro-intestinal carcinoma when used in modification of the Lemon-Foley method of administration. Toxicity is not a problem and therapy may be maintained for years as long as it is clinically effective. Patients with liver metastasis at the time of primary surgery demonstrate only improvements symptomatically with no evidence of tumor regression. In patients free of liver metastasis, palliation is often prolonged and striking. A long term double blind study of patients with Duke's C lesions treated by multiple courses of 14 days 5FU in addition to weekly bolus therapy is necessary. Second look surgery and radiation therapy can then be used more effectively.

REFERENCE

Engelfinger, F.J., 1966, Controversies in Internal Medicine,
    W. B. Saunders, Philadelphia, London, Toronto

SECTION II: ANTINEOPLASTIC EFFECTS OF RADIATION THERAPY AND
CHEMOTHRAPY BY CONTINUOUS UNFUSION

    A. Principles and Therapeutic Applications

THEORETICAL BASIS AND CLINICAL APPLICATIONS OF

5-FLUOROURACIL AS A RADIOSENSITIZER

John E. Byfield

Daniel Freeman Hospital

Inglewood, California, 90301, USA

INTRODUCTION

During the past 5 years several institutions have combined the anti-metabolite 5-Fluorouracil (5-FU) and radiation in the treatment of various human cancers. The results of these studies are thus far uniformly promising but as yet no randomized investigations have been conducted. The stimulus to these pilot studies came from two disparate sources. The first was the empiric observation by Nigro et al (23) that an infusion of 5-FU (combined with Porfiromycin or Mitomycin C), coupled in part with radiation, led to rapid pre-operative regression of squamous anal cancer. These results led to an application of this type of pre-operative program to other cancers (15, 17, 20, 26).

RADIOBIOLOGICAL BASIS OF 5-FU RADIOSENSITIZATION

The second stimulus has been the identification of the scheduling requisites for the application of 5-FU as a true radiosensitizer (RS) of human tumors (7). 5-FU has been known to be a RS for about two decades. The RS properties of 5-FU were first demonstrated in tissue culture by Bagshaw soon after its introduction (2). Mouse leukemia cells were subsequently shown to be strikingly RS in vivo (29). These original pre-clinical studies stimulated a wealth of early clinical trials using various combinations of 5-FU and external beam radiation. Some of these early studies suggested clinical benefit, others did not. All employed bolus 5-FU, usually in some variant of the original Wisconsin regimen in which daily, bolus 5-FU to toxicity (usually myelosuppression) was given followed by lower dose maintenance bolus 5-FU.

However, it was eventually shown in the author's laboratory (7) that bolus 5-FU cannot RS because of the pharmacological requirements of the RS phenomenon. Basically, the RS state is a cellular condition that develops gradually in the 24 hours or more after a radiation exposure. Moreover, it occurs only if the cell is exposed to an adequate concentration of 5-FU. Exposure of the cells to 5-FU prior to radiation has no sensitizing effect although additive toxicity (and tumor response) may occur. At the cellular level it appears likely that the radiation exposure may actually be sensitizing the cells to 5-FU (7,14).

Since tissues with slow turn-over times (nerve sheaths, vascular tissue, connective tissue etc.) are not affected by 5-FU, most late effects of radiation are not enhanced. Therefore full doses of radiation may be used provided appropriate adjustments in fractionation are employed to prevent serious epithelial damage that may itself lead to adverse late radiation effects.

We have also demonstrated that RS by 5-FU is not an arcane cellular event. It requires a substantial cytotoxic effect of the drug itself to become significant (7,14). In tissue culture this means RS is found only when there is sufficient drug (on a concentration X time basis) to produce a kill of about 30% of the cells without added x-rays (7). In terms of clinical applications, RS can probably always be anticipated if enough infused 5-FU is given to achieve a partial response (i.e. a PR without any added x-rays). This supposition will hold provided the tumor is not intrinsically radioresistant since in such cases RS by 5-FU seems unlikely. Clinically, this makes 5-FU RS therapy most useful in epithelial tissues which are sensitive both to 5-FU and radiation. Since the quantitative cytotoxicity of 5-FU is usually not first estimated in a clinical trial the administration of each 5-FU infusion to clinical toxicity is theoretically desirable (7,12).

The exact reasons that bolus-equivalent 5-FU does not RS are not known. It may well stem from the mode of action of the drug in some cells. Infused 5-FU may poison some types of tissues (and their derivative cancers) through its accumulation in cellular RNA rather than the more commonly proposed mechanism of inhibition of thymidylate synthetase (1,14,18). This interpretation is consistent with both the temporal sequence for RS (drug after x-ray) and the relatively slow development of RS in irradiated cells.

CLINICAL SCHEDULING REQUISITES OF 5-FU RS

Once it was understood that RS required a significant period of time to develop, it became apparent that only slowly infused 5-FU could RS. This stems from the short half-life of 5-FU in the bloodstream. When 5-FU is administered as a bolus injection its half life in the blood is only about 10 minutes with most of the drug being rapidly degraded by the liver (22). In order to create the conditions required for RS (a reasonably constant exposure to extra-cellular 5-FU for about 24 hours after each radiation exposure), it is necessary to constantly renew the internal supply of drug. Currently, this can only be done adequately by a slow, continuous infusion. Accordingly, the clinical toxicity of slowly infused becomes quite relevant.

The importance of scheduling in the normal tissue sensitivity to 5-FU was first pointed out by Seifert et al (25). They showed that the limiting toxicity of 5-FU reversed itself when the schedule of administration was changed from bolus to a 5-day infusion. During bolus 5-FU therapy bone marrow toxicity is almost always limiting, and is sometimes lethal (25). However, when a 5-FU infusion is given for 4-5 days to toxicity, stomatitis becomes dose-limiting and marrow suppression is usually very mild. This inversion effect on the dose-limiting tissue is relative rather than absolute. Thus the incidence and severity of each toxic reaction increases as the dose rate increases

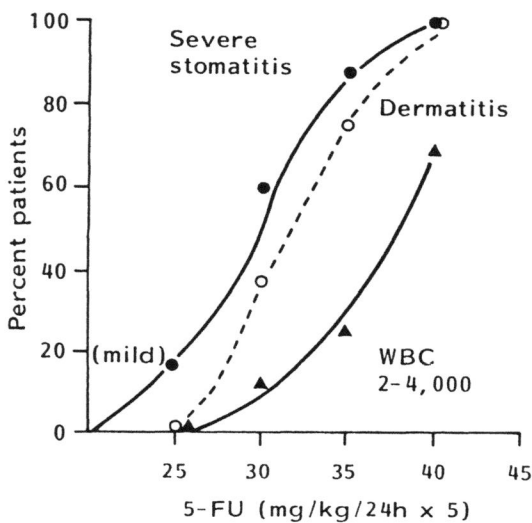

Fig. One. Dose-response relationship for the development of epithelial and marrow side effects in patients receiving a 5-day 5-FU infusion (data from ref. 25). The slopes of the curves are similar but the threshold varies, marrow having a higher threshold than mucosa.

for 5-day 5-FU infusions. This is illustrated in Fig. One. where the incidence of toxicity (from Seifert et al's data, ref. 25) for 120 hour infusions is plotted versus the daily 5-FU dose. It can be seen that the reaction generated in all three tissues evidencing toxicity (i.e. mucosa, skin, and marrow) increases with 5-FU dose. However, both the skin and the marrow have significantly higher thresholds than the oral mucosa. This small difference in threshold is, nevertheless, sufficient to convert a potentially lethal schedule (bolus 5-FU to toxicity) to a program that is clinically tolerable and safe (a 5-day continuous 5-FU infusion to toxicity). This aspect is quite important for the combination of the drug with radiation, especially for pelvic cancers.

These observations illustrate the three central requisites for applying 5-FU as a RS in man: (a) 5-FU RS is a post-radiation phenomenon and requires that 5-FU be available to the cells for 24 hours after each radiation exposure; this in turn requires a slow, constant infusion; (b) the more drug that is present (i.e. the higher the concentration for any given time period), the greater the likelihood of tumor response to the drug and therefore the greater chance that RS will develop; (c) there is also the reasonable possibility that many tumors, particularly those derived from epithelial tissues, are more likely to respond to slowly infused 5-FU than to bolus 5-FU (14,25).

These three pre-requisites are inter-dependent but all must be recognized in the development of ideal clinical regimens. At this stage of research there is no reason to believe that the optimal C x t factor for all human tumors types will have a constant value. Since the turnover times of both normal tissues and cancers can vary significantly, the optimal C x t factors for inducing 5-FU RS in different tumor sites probably also vary.

STATUS OF CURRENT CLINICAL STUDIES

To date the combination of infused 5-FU and radiation has been applied in several cancers including esophagus, head and neck, lung (13), colo-rectal, bladder, and squamous anal carcinoma. A related drug, 5-Fluorouridine deoxyriboside (5-FUdR), has been used with radiation in metastatic colon cancer to the liver (3,4) but will not be discussed further here because 5-FUdR does not RS. Several of these applications are covered in greater detail in other presentations in this volume. In only one cancer is there sufficient long-term experience to be able to state that the results of this approach are better than any other available regimen. This is its application in squamous anal cancer (reviewed in this volume and in ref. 15). In anal cancer the current challenge is to determine which patients with smaller tumors do not need this combined modality approach. For all other malignancies the available results are still limited. In many cases the inclusion of a second chemotherapeutic agent makes it difficult to interpret whether or not the 5-FU and the radiation component have been "optimized". Nevertheless, in all cases the results are encouraging enough to warrant further trials.

Head and Neck Cancer

The data available for this group of cancers is limited but of interest since it is in the head and neck region where the primary limiting toxicity of (local) radiation and infused 5-FU overlap; both modalities can induce severe but reversible oral mucositis (stomatitis). Stomatitis appears to be due to an accumulation of the drug within the cells of the limiting tissues. This is shown in Fig. 2 where the total dose of 5-FU needed to develop stomatitis is plotted against the daily dose rate. The data are derived from two phase I trials that studied the toxicity to 5-day and long-term continuous infusions of 5-FU respectively (21,25). It can be seen that a reasonably constant plateau for the total 5-FU dose needed to achieve stomatitis was found. This phenomenon is consistent with the existence of a highly efficient drug removal and accumulation mechanism (see under pharmacology, below), most probably entrapment of the 5-FU as a toxic product in cellular RNA (1,14,18).

The results of the existing head and neck studies, still mainly available in abstract form, are shown in Table 1. Our own group recently published the first phase I-11 trial of 5-day infused 5-FU combined with

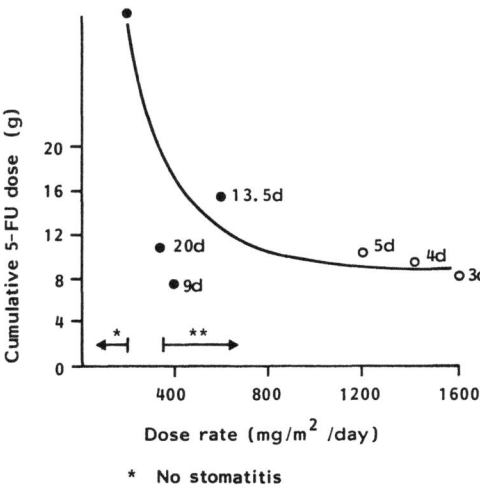

Fig. Two. Relationship between the total dose of 5-FU administered and the daily dose rate. Circles indicate the total duration of the average infusion for each dose rate. Data from references 21 and 25.

radiation in advanced squamous head and neck cancer (12). We have also recently reported some limited data on 72-hour infused 5-FU in this patient group (9). The first trial consisted of repeating cycles of 5-day infused 5-FU coupled with radiation at a constant dose rate (250 rads/fraction given on each of the first 4 days of each treatment cycle). The goal of the trial was to evaluate both toxicity and response as a function of 5-FU dose rate. The 5-FU dose rate was escalated in a phase I format.

As expected the commonest toxicity (and invariably the dose-limiting toxicity) was stomatitis. Since both infused 5-FU and head and neck radiation have stomatitis/pharyngitis as their major toxic manifestation this was to be expected. No patient could tolerate more than 25 mg/kg/day 5-FU (for each 5 day treatment cycle) for very long (2 cycles was the maximum). Even at 25 mg/kg/24 hours about 25% of patients required dose reduction to a lower level. On the other hand there was a clear-cut threshold for this toxicity since only mild stomatitis was induced at the lowest dose tested (20 mg/kg/day). All 20 mg/kg/day patients completed therapy without requiring dose reduction, despite the added radiation. However, there was also considerably reduced patient survival seen with 20 mg/kg as opposed to higher doses suggesting a therapeutic dose response relationship. In general, toxicity, response, and survival appeared linked in these results, as predicted by the theoretical basis of this treatment (12). The study showed clearly that the combination could be administered in a reasonably predictable fashion, even in head and neck cancer, with a suggestion of an improved response rate and survival benefit.

Table One. Results of infusional 5-FU and radiation in advanced squamous head and neck carcinoma.

Senior Author	# Pts.	CR%	Survival Median	2 Years	Reference
Showell	27	60	2 years	70%	24
Hahn	44	60 (s)	-	-	17
Byfield	18	75	2 years	70%	12

(s) - tumor free specimen at surgery.

There have been two other studies of similar combinations in head and neck cancer, each with the addition of another drug. Showell et al reported a study in which infused 5-FU and radiation was combined with platinum in advanced head and neck cancer (24). They found a high response rate and what they felt were clearly superior survival results compared to their previous studies.

Hahn et al combined Mitomycin C with infused 5-FU and external beam radiation in 44 patients with advanced head and neck cancer (17). Their program was a split course treatment using 3,000 rads in 3 weeks with the drugs being given "up-front", i.e. a bolus of Mitomycin C (10 mg/m2) on day one and a day 1-4 day infusion of 5-FU (1.0 g/m2). Most patients had subsequent surgery with a 62% negative primary specimen and a 33% negative neck specimen. The overall clinical complete response rate was 60%. They felt their results were promising compared to their historical experience with this bad prognosis patient group.

As noted above, we have also studied in a limited fashion the tolerance of patients with various tumors (including head and neck cancer) to 72-hour infused 5-FU (8,9). This latter study was done in order to see if it was possible to reduce the duration of each infusion which would be easier on the patient and more cost-effective if equally efficacious. However, we encountered significant 5-FU CNS toxicity during these high-dose, reduced duration exposures, and now feel that this aspect makes such short term infusions less desirable than infusions lasting 5 days or more (8,9).

Esophageal Cancer

Our first trial of 5-FU used specifically as a RS was in cancer of the esophagus. That trial used a dose of 20 mg/kg/24 hours 5-FU and a 5-day infusion duration (5). It also introduced the use of cyclical therapy with RS for the first time. Although primarily of an exploratory, phase I nature the study yielded a clinical complete response in almost all patients showing that the approach was both possible and useful in man. Noteworthy is the observation that the first patient so treated lived over 5 years free of disease before being lost to follow-up. Another patient in this initial series developed a carcinoma of the hypopharynx

Table Two: Results of infusional 5-FU and radiation in esophageal carcinoma.

Senior Author	# Pts.	CR%	Survival Median	2 Years	Reference
Byfield	6	83	22 mo*	30%	6
Keane	35	48	12 mo	28%	19
Leichman	21	25**	18 mo	24%	20

* survival data from entire UCSD experience only partly shown in ref. 6.
** defined as tumor-free after esophagectomy.

for which he underwent a second series of infusions with x-rays without substantial problems. This observation suggested that there is no demonstrable cumulative toxicity with the regimen.

There have been several subsequent studies using an approach similar to the one just described, usually with an added agent. These are summarized in Table 2. Some, such as the study from Wayne State, used an added agent (in this case cis-platinum) and used pre-operative radiation doses. In that study (by Leichman et al, 20), it was found that about one-third of the patients achieved a surgically-proven complete response. Median survival of the entire group was 18 months while those whose response was satisfactory enough to go to surgery was 24 months. All long term survivors were tumor-free at resection (including lymph nodes). 60% of these pathologic CR patients have lived at least two years. These results clearly relate tumor clearance (that is a complete response) with long-term survival. The authors also concluded that their results were better than they had previously achieved using a Mitomycin C containing regimen.

A similar report was recently issued by the workers at the Princess Margaret Hospital (19) but in their study no surgery was employed and the second agent used was Mitomycin C. In this study by Keane et al two different schedules of 5-FU were evaluated, in one case giving one RS cycle and in the other two such cycles (in a split-course approach after a four week rest). The 5-FU dose was 1.0 g/m2. Continuous radiation seemed better than split-course. Survival was 48% at one year and 28% at two years. The local control rate at two years was 48%. The authors concluded that these results were clearly superior to their historical results from radiation alone.

The studies cited above are the major current results of infused 5-FU used with coincident radiation in esophageal carcinoma. The reader is referred elsewhere in these proceedings for a more detailed summary. However, it should be noted that none of the existing studies indicate that this treatment approach has been used to maximum advantage. In none of these studies was a maximum dose of 5-FU used in every patient or even in every cycle in any given patient. Since the RS phenomenon is clearly 5-FU dose-dependent, it seems likely that a higher CR rate could be achieved with further modifications. The data of Leichman et al illustrate the relationship of a CR with survival. No such relationship would be apparent if death from metastatic disease was the sole important parameter in esophageal carcinoma.

Metastatic Colo-rectal Cancer in the Liver

We have recently reported our results of treating metastatic colo-rectal cancer to the liver with a combined modality regimen of 5-FUdR and whole liver radiation (3,4). Those results are fully published and will not be reviewed in detail here. The major observations made were: (a) this approach is feasible and can lead to patients with a CR; (b) unless delivered before liver function derangement is severe (greater than twice normal LFT's), no benefit derived; (c) this approach is limited by the development of toxic hepato-biliary damage; (d) almost all patients who live long enough will develop and die from metastatic disease outside the liver.

Cervix Cancer and Radiation of the Lower Bowel

There has been one trial thus far of this approach in advanced cervix cancer. That study was again modelled after the anal cancer programs but omitted surgery. Thomas et al used split course radiation therapy; a 72-96 hour 5-FU infusion was added to the radiation at treatment inception and again at the resumption of radiation treatment after the mid-course break (28). As in the anal programs, Mitomycin C was given as a single bolus each time 5-FU was started. In stage III B patients rapid tumor resolution was found and a complete response rate of 74% and one year control rate of 70% was found, compared with an expected control rate of 43% from historical controls. Patients with recurrent disease also appeared improved. The authors felt a randomized trial was indicated based on these results (28). The program is also important because the radiation portals involved lower gastro-intestinal radiation. Acute radiation toxicity did not appear much different than expected from radiation alone. Some adverse late effects on the bowel were also found, but again at an incidence level similar to radiation alone (28).

Squamous Anal Cancer

The results of the various programs used against squamous anal cancer have been reviewed at length elsewhere (15) and are further summarized in these proceedings by Cummings. It is important to recognize that this approach has been quite successful in this rather uncommon squamous cancer. In the series from Toronto (15) and San Diego (6) no surgery was required to achieve long term, tumor-free survival in almost all patients. These results clearly show the effectiveness of this regimen against squamous malignancy. An example of shown in Fig. Three.

Rectal Carcinoma

Experience with infused 5-FU used a RS is more limited in rectal cancer than in squamous anal cancer despite the more common occurence of the former. Recently, Sischy up-dated his results of using this approach (along with Mitomycin C) in marginally operable rectal carcinoma (26). He found (using 1.0 g/m2 5-FU) that 31% of the resected specimens contained no carcinoma while only 24% of peri-rectal nodes contained cancer

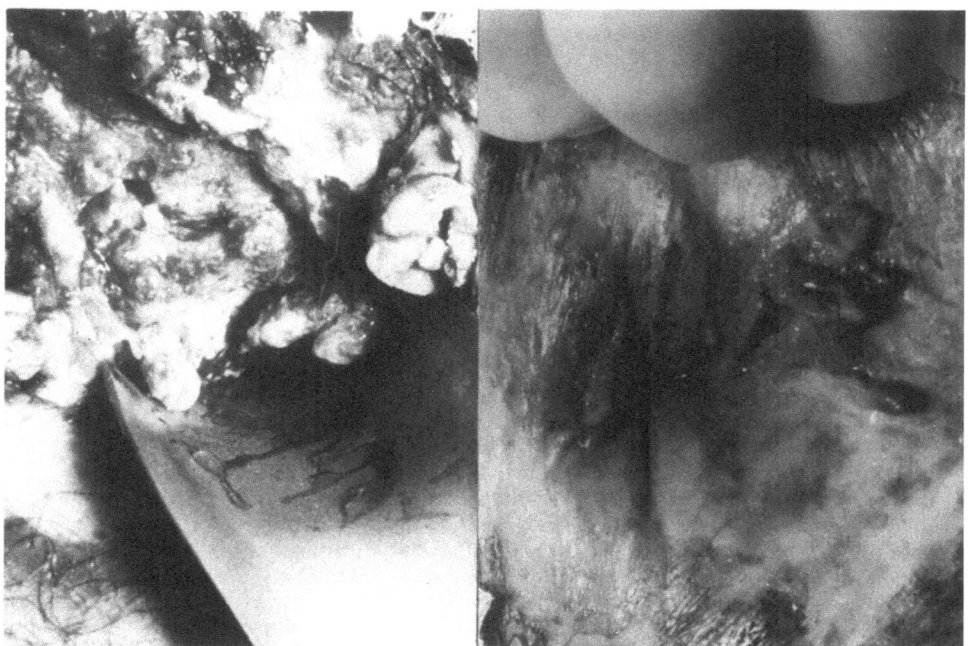

Figure Three. Response of an advanced squamous anal cancer after two treatment cycles (2,000 rads) combined with two 5-day infusions of 5-FU at 25 mg/kg/day. Left, prior to therapy. Right, after two cycles of RS 5-FU (2,000 rads).

compared with 53% of their historical controls. Local control in patients eligible for such 5-year evaluation was 84.9%. The treatment proved much less effective against recurrent disease.

Our own experience is more limited than that of Sischy et al but of a similar trend. We have also been able to induce CR but a rigorous evaluation has yet to be carried out. However, it is clear to those with experience in this cancer that the adenocarcinomas of the rectum are less responsive than their close neighbors, squamous anal carcinoma. Since these cancers originate within a few centimeters of each other, the origins of this difference is an obvious topic for future biochemical and cell kinetic investigations.

Pharmacology of Infused 5-Fluorouracil

When humans are infused with 5-FU one encounters an unusual phenomenon. The clearance of the drug is non-linear (16,22). This means that the serum 5-FU levels obtained are not a simple function of the amount of drug administered. At low infusion rates (anything significantly below 15 mg/kg/day) little to no drug is found in the plasma (8,9,16). Rather, the drug "disappears". We have proposed that this is due to its total removal from the blood stream at a single pass during low dose infusions. In this model the drug is actually accumulating within cycling cells rather than being degraded by the liver or other organs. This seems likely since the maximum clearance is essentially the cardiac output at a daily dose of just over 10 mg/kg/24 hours (8,9).

As the infusion rate increases above 10-15 mg/kg/day, 5-FU thereafter appears in the blood on a linear dose-concentration basis (8,9). Since it is quite clear that the toxicity of 5-FU relates to both the extra-cellular 5-FU level and the duration of tumor cell exposure to the drug (14), it is apparent that both the dose rate of an infusion and its duration are critical in the development of 5-FU RS. Future controlled trials should take cognizance of these features if an absolute evaluation of the potential of 5-FU as a clinical RS is to be obtained.

CONCLUSIONS

The primary modalities of therapy in the above programs are infused 5-FU and radiation. In most cases, either Mitomycin C or cis-platinum has been added but their exact role is unclear. In none of this author's studies were either Mitomycin C or cis-platinum ever employed yet equivalent results were obtained. This is true of head and neck, esophageal and anal cancers. It is the author's belief that the major components of this approach are the combination of a 5-FU infusion and radiation. Therefore a full understanding of the optimal combination of 5-FU and radiation in individual tumor types appears desirable.

The application of infused 5-FU and radiation clearly appears useful in man. Neither the ideal infusion duration nor the most appropriate radiation fractionation scheme has yet been established. Against squamous cancers 5-day infusions (at 25 mg/kg/24 hours) coupled with conventional radiation (200 rads a day for 5 days) seems suitable in most patients. Patients with severe hepatic derangement, including extensive metastatic disease, need 5-FU dose reduction (16). At the clinical level the induction of mild 5-FU toxicity (grade 1-2 stomatitis) is believed to be desirable. This should occur just following each treatment cycle and should be (just) healed prior to resumption of therapy. Accordingly, cyclical treatment is needed with resolution of all side effects during each rest period (5,6,12,13). The author does not believe that the total dose of radiation need be compromised. Future clinical trials should be devoted to further defining the use of this combination in various cancers, to the development of satisfactory oral 5-FU pro-drugs (10,11,27), and to the study of modulators capable of enhancing the sensitivity of tumors currently known to be resistant to 5-FU.

The properly applied use of 5-FU RS regimens appears on the verge of offering significant advances, especially in the treatment of squamous and squamous-like cancers. The latter includes transitional bladder carcinoma as shown elsewhere in this volume. The author also has limited but promising data in locally advanced breast cancer where 5-FU has always played a significant role. Since many forms of mutilating cancer surgery are necessitated because of the failure of conservative therapies for squamous cancers and squamous-like cancers (laryngectomy, abdominal-perineal resection, esophagectomy, vulvectomy, penectomy, cystectomy etc) there is an added and urgent reason for exploring this approach further.

REFERENCES

1. Ardalan, B., Cooney, D., and MacDonald, J.S. Physiological and pharmacological determinants of sensitivity and resistance to 5-Fluorouracil in lower animals and man. Adv. Pharmacol. Chemother. 17: 289-320, 1980.

2. Bagshaw, M.A. Possible role of potentiators in radiation therapy. Amer. J. Roentgenol. 85: 822-833, 1961.

3. Barone, R.M., Byfield, J.E., Goldfarb, P.B., Frankel, S.S., Ginn, C., Greer, S., and Callipari, F.B. Intra-arterial chemotherapy using an implantable infusion pump and liver irradiation for the treatment of hepatic metastases. Cancer 50: 850-862, 1982.

4. Byfield, J.E., Barone, R.M., Frankel, S.S., and Sharp, T.R. Treatment with combined intra-arterial 5-FUdR infusion and whole liver radiation for colon carcinoma metastatic to the liver. Amer. J. Clin. Oncol. 7: 319-325, 1984.

5. Byfield, J.E., Barone, R.M., Mendelsohn, J., Frankel, S.S., Quinol, L., Sharp, T.R., and Seagren, S.L. Infusional 5-Fluorouracil and x-ray therapy for non-resectable esophageal carcinoma. Cancer 22: 376-382, 1979.

6. Byfield, J.E., Barone, R.M., Sharp, T.R., and Frankel, S.S. Conservative management of squamous anal cancer by cyclical 5-Fluorouracil infusion alone and x-ray therapy. Cancer Treatment Rpts. 67: 709-712, 1983.

7. Byfield, J.E., Calabro-Jones, P., Klisak, I., and Kulhanian, F. Pharmacologic requirements for obtaining sensitization of human tumors cells in vitro to combined 5-Fluorouracil or Ftorafur and x-rays. Inter. J. Radiation Oncol. Biol. Phys. 8:1923-1933, 1982.

8. Byfield, J.E., Frankel, S.S., Sharp, T.R., Hornbeck, C.L., and Callipari, F. Phase I and pharmacologic study of 72-hour infused 5-Fluorouracil and hyperfractionated cyclical radiation. Inter. J. Radiation Oncol. Biol. Phys. 11: 791- 800, 1985.

9. Byfield, J.E., Frankel, S.S., Hornbeck, C.L., Sharp, T.R., and Callipari, F.B. Phase I and pharmacologic study of 72-hour infused 5-Fluorouracil in man. Amer. J. Clin. Oncol. (in press).

10. Byfield, J.E., Hornbeck, C.L., Frankel, S.S., Sharp, T.R., and Griffiths, J.C. Relevance of the pharmacology of oral Tegafur to its use as a 5-Fluorouracil pro-drug. Cancer Treatment Rpts. (in press).

11. Byfield, J.E., Sharp, T.R., Hornbeck, C.L., Frankel, S.S., Floyd, R.A., and Griffiths, J. Phase I and pharmacologic study of oral Ftorafur and x-ray therapy in advanced gastro-intestinal cancer. Inter. J. Radiation Oncol. Biol. Phys. 11: 597-602, 1985.

12. Byfield, J.E., Sharp, T.R., Tang, S., Frankel, S.S., and Callipari, F. Phase I and II trial of cyclical 5-day infused 5-Fluorouracil and coincident radiation in advanced cancer of the head and neck. J. Clin. Oncol. 2: 406-413, 1984.

13. Byfield, J.E., Stanton, W., Sharp, T.R., Frankel, S.S., and Koziol, J.A. Phase I-II study of 120-hour infused 5-Fluorouracil and split-course radiation therapy for localized non-small cell lung cancer. Cancer Treatment Rpts. 67: 933-936, 1983.

14. Calabro-Jones, P.M., Byfield, J.E., Ward, J.F., and Sharp, T.R. Time-dose relationships for 5-Fluorouracil cytotoxicity against human epithelial cancer cells in vitro. Ca. Res. 42: 4413-4420, 1982.

15. Cummings, B.J. and Byfield, J.E. Anal Cancer, in Innovations in Radiation Oncology Research (H.R. Withers and L.J. Peters, eds.), Springer-Verlag (in press).

16. Floyd, R.A., Hornbeck, C.L., Byfield, J.E., Griffiths, J.C., and Frankel, S.S. Clearance of continuously infused 5-Fluorouracil in adults having lung or gastro-intestinal carcinoma with or without hepatic metastatses. Drug. Intell. Clin. Pharm. 16: 665-667, 1982.

17. Hahn, S.S., Kim, J.-A., and Constable, W.C. Concomitant chemotherapy and radiotherapy for advanced squamous cell carcinoma of head and neck. Inter. J. Radiation Oncol. Biol. Phys. 10: (suppl 2), p. 191, 1984.

18. Heidelberger, C. Fluorinated pyrimidines and their nucleosides. In, Antineoplastic and Immuno-Suppressive Agents, vol. 2 (A.C. Sartorelli and D.G. Johns, eds.),pp. 193-231, Springer-Verlag, New York, 1975.

19. Keane, T.J., Harwood, A.R., Rider, W.D., Cummings, B.J., and Thomas, G.M. Concomitant radiation and chemotherapy for squamous cell carcinoma (SCC) esophagus. Inter. J. Radiation Oncol. Biol. Phys. 10 (suppl 2): 89 (abs), 1984.

20. Leichman, L., Steiger, Z., Seydel, H.G., Dindogru, A., Kinzie, J., Toben, S., MacKenzie, G., and Shell, J. Preoperative chemotherapy and radiation therapy for patients with cancer of the esophagus: a potentially curative approach. J. Clin. Oncol. 2: 75-79, 1984.

21. Lokich, J., Bothe, A., Jr., Fine, N., and Perri, J. Phase I study of protracted venous infusion of 5-Fluorouracil. Cancer 48: 2565-2568, 1981.

22. Myers, C.E. The pharmacology of the fluoropyrimidines. Pharmacol. Rev. 33: 1-15, 1981.

23. Nigro, N.D., Vaitkevicius, V.K., and Considine, B. Combined therapy for cancer of the anal canal: a preliminary report. Dis. Colon Rectum 17: 354-356, 1974.

24. Showell, J.L., Murthy, A.K., Hutchinson, L.D., Caldarelli, D.E., and Taylor, S.S. IV. Synchronous radiation therapy and cis-platin-5-FU chemotherapy in advanced head and neck cancer. Proc. 2nd Europ. Conf. Clin. Oncol. (abs), 162, 1983.

25. Seifert, P., Baker, L.H., Reed, M.L., and Vaitkevicius, V.K. Comparison of continuously infused 5-Fluorouracil with bolus injection treatment of patients with colo-rectal adenocarcinomas. Cancer 36: 123-128, 1975.

26. Sischy, B., Qazi, R., and Hinson, E.J. A pilot study of concurrent radiation, Mitomycin C, and 5-FU in marginally operable carcinomas of the rectum. Inter. J. Radiation Oncol., Biol., Phys. 10 (suppl. 2): 91 (abs), 1984.

27. Tang, S.G., Hornbeck, C.L., and Byfield, J.E. The potential enhancement of anti-tumor effect of Ftorafur by co-administration of uracil. Inter. J. Radiation Oncol. Biol. Phys. 10: 1697-1690, 1984.

28. Thomas, G., Dembo, A., Beale, F., Bean, H, Bush, R., Herman, J., Pringle, J., Rawlings, G., Sturgeon, J., Fine, S., and Black, B. Concurrent radiation, Mitomycin C, and 5-Fluoro uracil in poor prognosis carcinoma of cervix: Preliminary results of a phase I-II study. Inter. J. Radiation Oncol., Biol., Phys. 10: 1785-1790, 1984.

29. Vietti, T., Eggerding, F., and Valeriote, F. Combined effect of x-radiation and 5-Fluorouracil on survival of transplanted leukemic cells. J. Natl. Ca. Inst. 47: 865-870, 1971.

TREATMENT OF HEPATIC METASTASES FROM GASTROINTESTINAL PRIMARIES WITH SPLIT COURSE RADIATION THERAPY AND CONCOMITANT INFUSION 5-FLUOROURACIL

M. Rotman[1], J. Bhutiani[4], A. Kuruvilla[1], K. Choi[1],
J. Rosenthal[2], A. Braverman[2], and J. Marti[3]

SUNY, Downstate Medical Center, [1]Dept. of Radiation Oncology
[2]Medicine, [3]Surgery, 450 Clarkson Avenue, Brooklyn, New York
Winter Haven Hospital, [4]Dept. of Rad. Onc. Winter Haven, Fla.

INTRODUCTION

Metastases are detected in the liver in almost one third of cancer patients autopsied[16] and a 60-70% incidence[3] is not unusual for metastatic large bowel cancer. Taylor[21] reported progressive hepatic insufficiency as a cause of death in 25% patients who had radical resection for colorectal cancer.

Overt metastatic liver disease is currently treated with systemic chemotherapy, intraarterial or portal vein infusion of cytotoxic agents, hepatic artery ligation or embolization and hepatic irradiation with or without concomitant bolus chemotherapy. Surgical resection of solitary metastases has been offered in few centers, for even fewer highly selected patients.

Results of single modality treatment invariably showed a high morbidity and limited success due to problems of tumor burden, while combined modality therapy with its ability to increase cell kill was hindered by hepatic intolerance. In an attempt to improve the therapeutic ratio we initiated in 1980 a study of split course radiation with concomitant intravenous 5-fluorouracil for liver metastases from gastrointestinal malignancies at Downstate Medical Center. It is hoped that if such treatment shows significant prolongation of survival, with acceptable morbidity, its application as an adjuvant for patients at high risk of hematogenous liver metastases can be entertained.

MATERIAL AND METHODS

Twenty patients with hepatic metastases from gastrointestinal primaries were treated between April 1980 to May 1983. Follow-up ranged from 4 weeks to 113 weeks. There were seven males and thirteen females, aged between 41 to 78 years with a median age of 62 years. Fifteen patients had a primary cancer in the colon, two each had their primaries in the rectum and gallbladder respectively and the esophagus was the site of origin in one. Eleven patients had liver involvement only, as a site of metastases while 9 patients had liver metastases with other sites of involvement as well. The extent of liver disease was evaluated by either radionuclide liver scans, computerized axial tomograms and liver function tests, including alkaline phosphatase, serum glutamic oxaloacetic transaminase, lactic acid dehydrogenase and serum bilirubin.

Patients had liver irradiation with concomitant infusion 5-fluorouracil on weeks 1,3 and 5 of treatment and no treatment was given on weeks 2 and 4. The total dose to the liver ranged from 2400 rads to 3150 rads and radiation was delivered in 200 rad fractions, 5 days/week. Concomitant intravenous 5-fluorouracil infusion chemotherapy, 25mg/kg/24 hours was infused continually over 120 hours.

During treatment, patients were examined weekly with regards to weight, symptomatic and objective response, complete blood counts and liver chemistries. Response was determined by decrease in the size of the liver, assessed clinically and radiologically and decrease in post treatment values of initially elevated liver function indices.

RESULTS

There were 15 responders and 5 non-responders. Survival was measured from the initiation of liver treatment. The non-responders had a mean survival of 13.6 weeks and a median survival of 12 weeks, while responders had a mean survival of 48.3 weeks and a median survival of 36 weeks. Patients with only liver metastases had a median survival of 34 weeks vs. 20 weeks for patients with multiple organ metastases. Overall median survival for all of the patients was 24 weeks with 3 of the responders still alive at 89, 92 and 113 weeks follow-up. These preliminary results compare favorably with the Radiation Therapy Oncology Group, radiation alone studys' median survival of 11 weeks.[4] All patients who had a demonstrable decrease in liver size, also showed a decrease in serum LDH. A similar correlation was also seen with serum SGOT, but the decrease was less predictable.

COMPLICATIONS

No severe complications have been encountered with this combined modality treatment to the liver.

Nausea was the most frequent side effect occuring in 10/20 patients, while vomiting attributed to treatment occured in only 1/20. Anorexia and lassitude were noted in almost all the patients but this was generally a presenting complaint which improved in all, except in the non responders whose general condition progressively deteriorated. Other G.I manifestation of toxicity observed were, oral mucositis in 6/20 patients while 4/20 were noted to have superficial G.I ulcerations, but neither of these were debilitating, and responded to conservative measures.

Hematological tolerance was acceptable as only 3 patients developed a WBC count of less than 2000 and one patient had a platelet count of 30,000. These instances of leukopenia and thrombocytopenia spontaneously corrected without necessitating transfusion. Thrombocytopenia, due to decreased platelet survival has been reported following liver irradiation in elderly patients.[15] The pathogenesis is uncertain but spontaneous recovery of the platelet count almost always occurs.

DISCUSSION

Heidelberger[10] et al in 1958, showed that ineffective doses of either irradiation or 5-fluorouracil could be made curative for certain rodent tumors, if both treatments were combined. Vermund[23] et al in 1964, performing in-vitro experiments, confirmed the in-vivo findings of Heidelberger[10] et al. Some light on the possible mechanism of action of radiation and 5-fluorouracil was explained by work in the laboratory and clinic by Vietti[24] et al. in 1971, and Byfield, et al.[5] namely that responses ranging from additivity to marked potentiation could be attained and that synergism was critically dependent on the temporal sequence of the two agents.

The effect when the drug was administered BEFORE radiation was interpreted as due to synchronization of the tumor cells by the 5-fluorouracil. In patients with colonic carcinoma with hepatic metastases, treated with 5-fluorouracil alone, response rates of 15-40% (50% with intraarterial 5-fluorouracil) have been observed, and this is attributed to the action of the drug in the S phase, inhibiting thymidylate synthetase, resulting in impaired Thymidylic acid formation an essential DNA precursor. This reduction of radioresistant S phase cells, leave a depleted population which are sensitive to x-rays.

The sensitizing effect of addition of 5-fluorouracil AFTER irradiation has been hypothesized to be due to inhibition of x-ray induced sublethal damage repair. This is very plausible, as 5-fluorouracil is known to inhibit RNA formation in addition to DNA inhibition. It is very likely that impairment of the RNA salvage pathways by the drug, which are known to be necessary to repair DNA damage, prevent repair of x-ray induced sublethal DNA damage in the metastatic tumor cells.

Seifert[18] et al in 1975 reported results of a prospective randomized trial demonstrating that chemotherapy administered as an infusion had certain advantages over bolus administration. This is thought to be because the synthetic phase of the tumor cell cycles vary from minutes to hours, and similarly the plasma half life of most cytotoxics have a short range when used as bolus, therefore, infusion overcomes the disadvantage of bolus administration by not only allowing prolonged exposure of the cytotoxic to the tumor cells, while additionally causing less hematopoetic suppression, the dose limiting target for most cytotoxics. Intravenous infusion allows the drug to be present before and after radiation for optimal radiosensitization.

As prolongation of continuous 5FU infusion over 5 days would lend itself to unacceptable toxicity, the patients are treated with split courses to allow for recovery of the affected normal tissues. In addition, it is hypothesized that the short split courses of radiation would allow for rapid regeneration of normal healthy cells without undue repopulation of the diseased and slowly growing adenocarcinoma metastatic to the liver. The gap would also allow for removal of necrotic tumor cells, the resultant tumor shrinkage facilitate reoxygenation and improve vascularity. This in turn would enable the previously hypoxic radioresistant tumor cells move into the oxic radiosensitive compartment and also enhance chemoperfusion of the lesions in subsequent courses following the rest period.

In 1924, Caisse and Warthin[7] were the first to refer to the usefulness of hepatic irradiation in treating patients with metastatic disease. However, they erroneously interpreted the autopsy findings in the liver of their 3 treated patients as they suggested the liver could tolerate high dose irradiation without adverse clinical effects. This impression of radioresistance remained entrenched until Ogata and associates[15] (1963) described radiation changes in the liver following doses of 1500 to 7000 rad. Ingold, Kaplan and Bagshaw[11], in 1973, reinforced the syndrome of Radiation Hepatitis, describing the clinical feature of hepatomegaly, rapid weight gain, ascites, accompanied by jaundice and deteriorating liver biochemistry. The clinical picture resembled that of the veno occlusive syndrome of BUDD CHIARI. Histologically the acute findings (within 2 months) consists of centrolobular hemorrhage, sinusoidal dilatation and congestion, with minimal adjacent hepatocellular atrophy. Late histological changes (after 3 months) are characterized by atrophy of centrolobular hepatic cords, minimal erythrocyte distension of the sinusoids and thickening of the central vein wall ranging from endothelial hyaline thickening, to complete venous occlusion.

The Stanford study demonstrated a dose response relationship and established the dogma of a "safe" dose in the range of 2500-3000 rad to the whole liver to be within hepatic tolerance. Tefft[22], et al., pointed out in 1970 the above could not be applied to the pediatric population and that pediatric liver tolerance would be exceeded, above 1200-1500 rad, especially with Actinomycin D or partial hepatic resection.

Despite the above wealth of experience and knowledge, the liver TTD-5 (Tissue tolerance dose, 5% lethality) is yet uncertain for various fractionation schemes (conventional hypo, accelerated, hyper or hybrids of these fractionation schemes) and the Radiation Therapy Oncology Group is currently attempting to answer some of these in a prospectively randomized fashion (RTOG Protocol 84-05).

Treatment with radiation therapy alone was investigated by the Radiation Therapy Oncology Group[4] (76-05). This was a non randomized pilot study which examined a variety of fractionation schemes used in the palliation of hepatic metastases. 2100 rads, 300 r/Fx x 7 was clinically well tolerated and apparently as effective in palliative symptoms as more prolonged schemes using smaller fraction sizes carried to a higher dose (3000 r, 2 Gy x 15 or 2520 r, 1.6 Gy x 16). The median survival of the 109 patients entered into that study was only 11 weeks, whereas, in this Downstate Medical Center series, patients treated with concomitant intravenous infusion 5-fluorouracil and radiation have an overall survival of 24 weeks.

Prasad, Lee and Hendrickson[17] in 1977 published their experience with 20 patients whose livers with metastatic disease were irradiated with an average dose of 2500 in 3-3.5 weeks. Palliation was achieved in 70% and a median survival of 4.5 months were noted for the entire group.

Experience with Systemic 5-fluorouracil alone for liver secondaries from solid G.I tumors result in an objective response of 15% lasting for 3 months or more but survival is unaffected.[6]

Attempts at selective intraarterial infusion for liver secondaries were first initiated in 1950. This is based on the observation that metastatic neoplasms derive their blood supply from the hepatic artery, while normal parenchyma obtains 80% of its vasculature from the portal vein. Systemic toxicity is also less as almost 80% of 5-fluorouracil is removed by the liver in its first pass. Sullivan[20] in 1959 introduced the method of intraarterial infusion using 5-fluorouracil deoxyriboside (FUdR) and showed a 59% response rate in metastases from colorectal primaries, for a minimum of 3 months.

In 1979, the Central Oncology Group[9] reported results of a prospective randomized clinical trial comparing treatment of colorectal liver metastases with intra arterial hepatic artery 5-fluorouracil infusion vs. systemic 5-fluorouracil. While this study suggested no significant difference between systemic and intra arterial infusion therapy (median survival time of 13 months vs. 15 months), a significantly higher complication rate was noted for the intra arterial infusion treated patients, compared with the patients receiving systemic chemotherapy. This study has been challenged as the numbers in each group were small, and apparently the intra arterial group received suboptimal therapy.

Experience with selective hepatic arterial infusion has accumulated with 5-fluorouracil and its deoxyriboside derivative (5-FUdR), individually or in combination with other drugs (mitomycin, Adriamycin) and irradiation. Major technical refinements have included the transition from temporary percutaneous arterial drug delivery system to permanently implanted subcutaneous arterial infusion pumps. However, to date results of a prospective controlled randomized trial comparing this approach with others are not yet available, and the influence of natural selection cannot be ruled out, as patients treated with these techniques have to be able to withstand a laparotomy for arterial catheterization.[2,6,12,13,14,19]

The occasional patient, with a small (less than 5cm) well differentiated lesion, occuring some years after initial diagnosis, may benefit by a wedge resection at the time of initial colonic resection, with low operative mortality.[1]

CONCLUSION

The treatment of hepatic metastases from G.I primaries has been reviewed. Reported here is our experience with split course liver irradiation with concomitant infusion 5-fluorouracil. The patients analyzed so far are representative of the natural history of the disease in that the metastases were present at diagnosis, resection of their primary tumor or within 2 years thereafter.

Our experience suggests that this treatment results in palliating the distressing symptoms in the majority of patients treated with a satisfactory prolongation of survival. With overt hepatic metastases, no form of currently employed therapy offers any hope of cure, unless truly solitary and surgically resectable. Combination of irradiation and 5-fluorouracil in this fashion has been shown to improve results when compared with either modality used alone. Though the exact mechanism of radiosensitization has yet to be determined, preferential tumor cell sensitization appears to be occuring. The sigmoid response curves of tumor control versus complications appear to have been shifted favorably providing greater tumor control, and the therapy is well tolerated and follow up so far not demonstrated any excessive or unacceptable acute toxicity.

Bearing the above observations in mind, an honest appraisal of our own selected and non randomized results, and scrutiny of the others reviewed, was made. It is clear that any definite and meaningful conclusion can only be obtained after a carefully matched group of patients, with adequate controls, are given optimal treatment in a prospective and randomized multi-institutional study. Only in this way can it be determined if the dose of 5-fluorouracil and radiation, and the sequencing employed, provide the optimum palliation and indeed does significantly prolong median survival, without any undesirable long-term sequelae.

The ultimate bold aim for the future may eventually lie in elective adjunctive liver irradiation with concomitant infusion 5-fluorouracil as a radiosensitizer to eradicate subclinical disease, in patients identified to have a high risk of hematogenous liver dissemination. This form of prophylactic therapy appears to be the only means of by which long-term survival and eventual cure may be attained.

REFERENCES

1. Adson, M.A., Van Heerden, J.A.: Major hepatic resections for metastatic colo-rectal cancer. **Ann. Surg., 191**:576-580, 1980.

2. Balch, C.M., Urist, M.M., Soong, S.J. et al.: A prospective Phase II clinical trial of continuous FUdR regional chemotherapy for colorectal metastases to the liver, using a totally implantable drug infusion pump. **Ann. Surg.**, 198, 5:567-573, 1983.
3. Berge, T.H. et al.: Carcinoma of the colon and rectum in a defined population. Acta Chir. Scand. (Suppl) **438**:1-86, 1973.
4. Borgelt, B.B., Gelber, R., Brady, L.W. et al.: The palliation of hepatic metastases: Results of the Radiation Therapy Oncology Group pilot study. **Int. J. Radiat. Onc. Biol. Physics,** 7:587-591, 1981.
5. Byfield, J.E., Calabro, Jones P., Klisak, I., Kulhanian, F.: Pharmacologic requirements for obtaining sensitization of human tumor cells in vitro to combined 5 flouracil or florafor and x-rays. **Int. J. Radiat. Onc. Biol. Physics,** 8:1923-1933, 1982.
6. Byfield, J.E., Barone, R.M., Frankel, S.S. et al.: Treatment with combined intraarterial 5-FUdR infusion and whole liver irradiation for colon carcinoma metastatic to the liver. **Am. J. Clin. Oncol. (CCT)** 7:319-325, 1984.
7. Caisse, T.J., Warthin, A.S.: The occurence of hepatic lesions in patients treated by intensive deep Roentgen irradiation. **AJR**, 12:27-46, 1924.
8. Foster, J.H.: Treatment of metastatic cancer to the liver. **Principles and Practice of Oncology**, Ed.: DeVita, V.T., Hellman, S., Rosenberg, S.A., J.B. Lippincott Co., Pg. 1557.
9. Grage, T.B., Vassilopoulos, P.P., Shingleton, W.W., Jubert, A.V., Elias, E.G., Aust, J.B., Moss, S.E.: Results of a prospective randomized study of hepatic artery infusion with 5-fluorouracil versus intravenous 5-fluorouracil in patients with hepatic metastases from colo-rectal cancer: A Central Oncology Group Study. **Sugery, 86**:550-555, 1979.
10. Heidelberger, C., Greisbach, L., Montag, B.J., Mooren, D., Cruz, D., Schnitzer, R.J., Grunberg, E.: Studies in fluorinated pyrimidines II. Effects on transplanted tumor. **Ca Res.**, 18:305-317, 1958.
11. Ingold, J.A., Reed, G.B., Kaplan, H.S., Bagshaw, M.A.: Radiation Hepatitis. **AJR,** 117:73-80, 1973.
12. Kemeny, N., Daly, J., Oderman, P. et al.: Hepatic artery pump infusion: Toxicity and results in patients with metastatic colorectal carcinoma. **J. Clin Oncol**: 2, 6:595-601, 1984.
13. Lokich, J., Kinsella, T., Perri, J. et al.: Hepatic radiation and intraarterial fluorinated pyrimidine therapy. **Cancer,** 48:2569-2574, 1981.
14. Neiderhuber, J.E., Ensminger, W., Gyves, J. et al.: Regional chemotherapy of colorectal cancer metastatic to the liver. **Cancer, 53**:1336-1343, 1984.
15. Ogata, K., Hizawa, K., Yoshida, M., Kitamuro, T., Agaki, G., Kagawa, K., and Fukuda, F.: **Tukushima J. Exp. Med.,** 9:240-251, 1963.
16. Phillips, R., Karnofsky, D., Hamilton, L. et al.: AJR 71:826-834, 1954.
17. Prasad, B., Lee, M.S., Hendrickson, F.R.: Irradiation of hepatic metastases. **Int. J. Radiat. Oncol. Biol. Physics,** 2:129-132, 1977.
18. Seifert, P., Baker, L.H., Reed, M.L.: Comparison of continuously infused 5-FU with bolus injection in treatment of patients with colorectal adenocarcinoma. **Cancer, 36**:123-128, 1975.
19. Stagg, R.J., Lewis, B.J., Friedman, M.A. et al.: Hepatic arterial chemotherapy for colorectal cancer metastatic to the liver. **Ann. Int. Med., 100**:736-743, 1984.
20. Sullivan, R.: Systemic and arterial infusion chemotherapy for metastatic liver cancer. **Int. J. Radiat. Oncol. Biol. Phys.,** 1:973-976, 1976.
21. Taylor, F.W.: Cancer of the Colon and Rectum. A study of routes of metastases and death. **Surgery,** 52:305-308, 1962.

COMBINED MODALITY THERAPY WITH 5-FLUOROURACIL, MITOMYCIN C AND RADIATION THERAPY FOR SQUAMOUS CELL CANCERS

B.J. Cummings, T.J. Keane, A.R. Harwood, and G.M. Thomas

Department of Radiation Oncology
The Princess Margaret Hospital
Toronto

The eradication of epidermoid cancers of the anal canal by an empirically developed program of concurrent radiation, 5-Fluorouracil (5FU), and Mitomycin C (MTC), was described by Nigro, Vaitkevicius and Considine in 1974. The rapid and complete tumor response obtained in the first three patients encouraged these investigators, and subsequently others, to evaluate the combination in patients with squamous cell carcinomas in different sites. It is perhaps fortuitous that epidermoid anal carcinomas were treated first, for the results in that site have generally been the most striking, and even then some authors have questioned whether similar results could not have been achieved with simpler regimens.

The three agents included in this treatment protocol have been studied as single agent treatment for many years. Radiation therapy is long established as a treatment for squamous cell carcinomas and frequently produces both local control and cure. 5-Fluorouracil, a fluoropyrimidine, produces responses in from 10 to 20 percent of patients with advanced squamous cell carcinomas in various sites. Mitomycin C, which acts principally as an alkylating agent, has been used in fewer studies than 5FU, but is also reported to produce responses in from 10 to 20 percent. There is much less information about the response of squamous cell carcinomas to combinations of 5FU and MTC. Michaelson et al (1983) recently reported a 59% response rate in patients with advanced epidermoid anal cancer. A trial conducted by the Southwest Oncology Group showed partial responses in 4 of 10 patients with disseminated esophageal cancer (Rosenberg et al, 1982). Most of these responses to either single agent or combined chemotherapy were partial and lasted only a few months.

The general principles and objectives of combining chemotherapeutic agents and radiation therapy have been reviewed extensively (Fu, 1985; Peckham et al, 1981). Although most investigators studying combinations of 5FU, MTC and radiation (FUMIR) have retained the concurrent administration of the agents as described in the original protocol, it is worth noting that, in general, the concurrent administration of drugs and radiation frequently enhances normal tissue effects, while tumor effects have a more variable dependance on the timing of chemotherapy and radiation (Fu, 1985). The justification for the concurrent use of 5FU and radiation in this protocol is based on both clinical and laboratory evidence. In the case of MTC, however, concurrent administration may not be necessary.

Prior to the introduction of FUMIR in Detroit (Nigro et al, 1974) there had been few studies of concurrent 5FU and radiation in squamous cell cancer. Gollin et al (1972) had reported improved local control and survival, but increased acute toxicity, in patients with advanced head and neck cancer treated by intravenous bolus injections of 5FU concurrently with radiation. Because the Detroit group had demonstrated that a continuous infusion of 5FU produced less myelosuppression than daily bolus injections (Seifert et al, 1975), they elected to study this schedule of 5FU administration in their FUMIR protocol. As Byfield discusses elsewhere in this volume, this may have been a fortunate choice, for in his laboratory studies with adenocarcinoma cell lines, any sensitizing interaction between radiation and 5FU was strongly dependent on concurrent use of the agents, and the interaction was also dependent on the maintenance of cytocidal levels of 5FU over a continuing period, in a way more analogous to continuous infusion than to bolus injection. Studies with MTC have demonstrated that at low concentrations, similar to those which can be achieved in humans by bolus injections, any effects of MTC and radiation on aerobic cells appear to be additive (Rockwell, 1982). The data on any possible interaction on anaerobic cells is more limited, but it has been demonstrated that, under experimental conditions, MTC is more toxic to cells that are hypoxic at the time of drug treatment than to well oxygenated cells (Teicher et al, 1981). It has been speculated that this differential cytotoxicity might be an advantage when MTC is combined with radiation in the treatment of tumors containing hypoxic cells, since such cells are relatively radioresistant. The minimum concentration of MTC necessary for true radiosensitization was calculated to be 4 ug/ml (Rockwell, 1982), about four times the peak plasma levels attained in humans by bolus injections of 10 to 30 mg (Reich, 1979). It is, of course, quite possible that none of these laboratory experiments is relevant to the clinical treatment of squamous cell carcinomas by fractionated courses of radiation combined with chemotherapy, but it is clear that further studies of this kind will be needed to enable more directed development of the empirical schedules currently employed in the clinic.

While the theoretical bases for the use of FUMIR are not yet resolved, many clinical studies have been reported. At the Princess Margaret Hospital (PMH), we have evaluated the response to FUMIR of squamous cell cancers in several sites. The main areas of interest have been anal and esophageal cancer, head and neck cancer, and uterine cervical cancer. The results achieved in these tumor sites at PMH, and elsewhere, are presented here.

PRINCESS MARGARET HOSPITAL TREATMENT SCHEDULES

The schedules used most frequently at PMH for the treatment of carcinomas of the anal canal, esophagus, and head and neck are summarized graphically in Figure 1. Schedule A involves external beam megavoltage radiation alone to a tumor dose of 5000 cGy in 20 fractions in 4 weeks. This radiation dose is retained in Schedule B, but in Schedule C the radiation course is split into two segments, each of 2500 cGy in 10 fractions in 2 weeks separated by a 4 week interval. In Schedule B, a single bolus intravenous injection of Mitomycin C, 10 mg/m^2, is given on day 1, and a 4 day continuous intravenous infusion of 5-Fluorouracil, 1000 mg/m^2/24 hours, is given on days 1 through 4. In the split course Schedule C, the same chemotherapy is given over the first 4 days of each of the radiation courses. For all primary tumor sites, the maximum dose of 5FU is restricted to 1500 mg/24 hours, and for esophageal and head and neck carcinomas the maximum dose of MTC is restricted to 15 mg. These dose restrictions were decided upon arbitrarily and other investigators have used different doses.

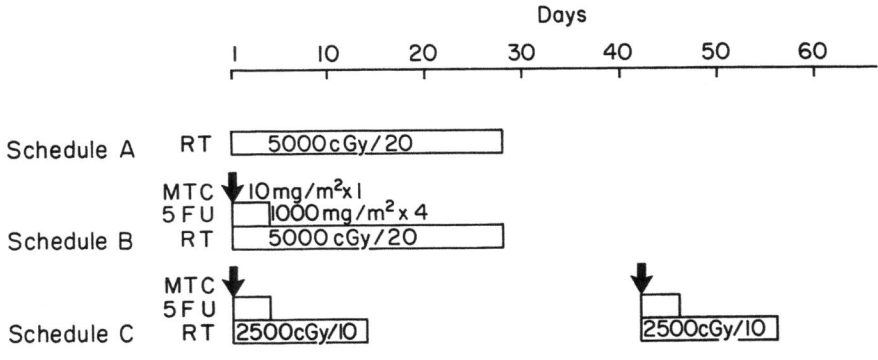

Figure 1. Princess Margaret Hospital Treatment Schedules. RT = Radiation Therapy; MTC = Mitomycin C; 5FU = 5-Fluorouracil. Figures in boxes represent radiation dose in number of centiGrays/number of fractions.

A completely different chemotherapy and radiation schedule was used for the patients with cervical cancer, and is described below.

ANAL CANAL CARCINOMA

Prior to the introduction of FUMIR in the management of anal canal carcinoma, the most frequently used treatment methods were radical surgical resection or radical radiation therapy (Cummings, 1982). Each of these methods produced 5 year survival rates ranging from 30 to 80 percent, and averaging about 50 percent, according to the pattern of patient referral. Direct comparison of the results of surgery and of radiation therapy has generally not been possible, mainly because of the lack of a commonly accepted staging system. Surgical resection in the form of abdominoperineal resection (APR) was used more often, principally because of concern about the potential toxicity of radiation therapy.

Two different approaches have been used to incorporate FUMIR into the management of anal cancer, according to whether a center had previously favored surgery or radiation therapy. In the early studies (Nigro et al, 1974) FUMIR was intended as preoperative adjuvant therapy, and moderate radiation doses of about 3000 cGy in 3 weeks were used. The protocol was subsequently modified so that APR was performed only if a biopsy from the site of the original tumor revealed residual carcinoma (Michaelson et al, 1983; Nigro, 1984; Sischy et al, 1982). A different approach was developed by investigators whose previous policy had been to use radical radiation doses and reserve APR for the patient with residual or recurrent carcinoma (Cummings et al, 1980; Papillon, 1982). In those centers, 5FU and MTC were added to programs of radical radiation therapy.

At the PMH, 5FU and MTC were first added to radiation in 1978 (Cummings et al, 1980, 1984). A single course of chemotherapy was given over the first 4 days of a radical course of radiation (Schedule B, Figure 1). This program produced unacceptable acute hematologic and enteric toxicity, and it was felt prudent to alter the treatment schedule to a split course protocol (Schedule C, Figure 1). Other groups who combined radical radiation doses with chemotherapy, also found that acute toxicity frequently necessitated a split course program (Sischy et al, 1982). The PMH results are shown in

Table 1. ANAL CANAL CARCINOMA

	Primary Tumor Control by RT	Severe Toxicity Acute[b]	Severe Toxicity Late[c]	Colostomy	5 year Actuarial Survival
RT alone	15/25[a]	4/25	3/25	8/22[d]	72%
FUMIR continuous	15/16[a]	9/16	3/16	3/16	74%
FUMIR split	13/14[a]	5/14	2/14	1/14	

[a] Successful surgical salvage 7/7 (RT); 1/1 (FUMIR-continuous); 0/1 (FUMIR-split)

[b] Number of patients

[c] Number of patients requiring surgical management of toxicity

[d] Excludes 3 patients who had colostomy prior to RT

Table 1, and are compared to those achieved with similar external radiation therapy doses in patients with comparable stage disease. The most notable difference is in the primary tumor control rates, where the two FUMIR programs resulted in control in 93 percent (28/30) compared to 60 percent (15/25) with radiation alone. We found that clinical examination was usually adequate to assess primary tumor response, and have not used biopsies following radiation or FUMIR except where residual cancer was suspected clinically. Salvage surgery was successful in seven patients in the radiation group so that the eventual pelvic control rates are not too dissimilar. However, the need for surgical management of residual cancer is reflected in the greater number of patients who required colostomies in the radiation group. It is noteworthy that all four colostomies in the FUMIR-treated patients were required for the management of late toxicity, due either to persistent ulceration at the primary tumor site, or to rectal or sigmoid damage. The various treatment schedules and the results approximately six weeks after the end of treatment in the PMH and Rochester series (Sischy et al, 1982) are shown in Figure 2. Late recurrence has not been a problem in the PMH series, in which only one tumor (in the split course FUMIR group) recurred locally three months after the end of treatment.

Our current technique involves irradiation of the primary anal tumor and inguinofemoral nodes and pelvic nodes to the level of the lower border of the sacroiliac joint for the first course of radiation, and reduction of the field to include only the anal area for the second course. The radiation schedule has been altered to two courses, each of 2400 cGy in 12 fractions in 2 ½ weeks, separated by 4 weeks. Preliminary analysis shows no change in either the primary tumor control rate or in acute toxicity with this lower daily fraction size which was adopted in the hope of reducing late toxicity.

The planned preoperative approach of Nigro and others has also proven successful. The schedules and results achieved in Detroit (Nigro, 1984) and in New York (Michaelson et al, 1983) are also shown in Figure 2 for comparison. Although designed as an adjuvant to abdominoperineal resection, the frequent finding of complete tumor regression led most centers to limit surgical resection to excision or biopsy of the residual scar (Michaelson et al, 1983; Nigro, 1984; Sischy et al, 1982). Only if carcinoma was identified in the biopsy was radical surgery performed. Nigro (1984) collected information on 104 patients either from his own group or by

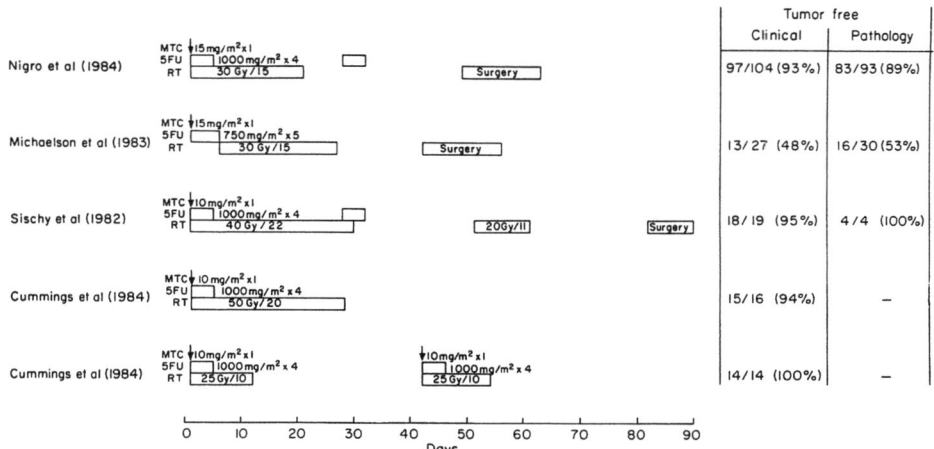

Figure 2. Selected treatment schedules for anal canal carcinoma. MTC = Mitomycin C; 5FU = 5-Fluorouracil; RT = Radiation Therapy. Figures in boxes represent radiation dose in number of Grays/number of fractions.

questionnaire, and reported that residual carcinoma was found on microscopy in only 3 of 86 patients who had no gross tumor following preoperative FUMIR (2 of 24 APR specimens; 1 of 62 biopsy specimens). In those who had a negative biopsy, 7 of 61 (11%) developed local recurrence within a year, but 4 of the 7 were alive three years after APR. Michaelson's experience (1983) was similar, and only 2 of 17 patients treated by limited resection following FUMIR developed locoregional recurrence. The acute toxicity of preoperative FUMIR using 3000 cGy in 15 fractions in 3 weeks to the pelvis was reported to be acceptable, with only 5 of 104 patients developing serious toxicity (Nigro, 1984). None of the reports of preoperative treatment commented on late toxicity.

Several questions arise from these studies - for example, what is the role of biopsy after treatment with FUMIR?, is concurrent chemotherapy and radiation more effective than sequential treatment?, is MTC necessary?, does FUMIR influence occult distant metastases?, and the basic question - is combined modality therapy better than radiation alone or radical surgery? While definitive answers are not possible, the widely reported results do permit some conclusions.

Biopsy following FUMIR would seem to be unnecessary unless residual cancer is suspected clinically. In the high radiation dose studies at PMH we have noted an excellent correlation between clinically complete tumor regression and long term control. We felt that selective biopsies could only sample a small part of the original tumor site at risk of harboring residual carcinoma, and wide excisions might not heal well following FUMIR with high radiation doses. In the lower radiation dose series of Nigro (1984) and Michaelson et al (1983), there was a very high level of correlation between complete clinical regression and the histological absence of tumor in the biopsy specimens (Figure 2). Even so, negative biopsies were still followed by local recurrence in 11 percent (7/61), so that a negative biopsy does not replace the need for regular and careful followup.

The theoretical grounds for using 5FU concurrently with radiation were outlined previously. In Figure 2 it can be seen that Michaelson et al

(1983) employed sequential chemotherapy and radiation rather than the concurrent programs used in the other schedules illustrated. Only 48 percent of their patients were rendered clinically tumor free compared to 93 percent reported by Nigro (1984). It is impossible to determine whether this difference was due to the alteration in scheduling, since other factors which differed included the use of two courses of 5FU by Nigro, the interval to surgery was shorter in the New York series, and the stage and size of the tumors treated may have been different. However, these results together with the experimental evidence of Byfield's studies (1982) suggest that concurrent 5FU and radiation may be preferable.

The role of MTC in these schedules has been questioned (Byfield et al, 1983), and the laboratory studies suggest that the concurrent administration of MTC and radiation may not be essential since cytotoxic effects appear to be additive rather than interactive (Rockwell, 1982). In a small series of patients, Byfield (1983) demonstrated that a schedule of repeated cycles of concurrent radiation (1000 cGy/4 fractions/4 days) and 5FU (25 mg/kg/24 hours by continuous intravenous infusion for 5 days) without MTC is also effective treatment for epidermoid anal carcinoma. Since only one or two doses of MTC are usually given in FUMIR schedules, the toxicity associated with cumulative doses of MTC, which may include myelosuppression, nephrotoxicity, pulmonary toxicity, and microangiopathic hemolytic anemia, has rarely been seen. While it is possible that MTC could be dropped from these protocols, none of the larger groups has yet reported the results of such a policy.

Occult distant metastases do not seem to have been affected by the one or two courses of chemotherapy in these schedules. The incidence of distant metastases which appear late while the primary tumor remains controlled is about 10 percent to 15 percent whether initial treatment is radical surgery, radiation, or combined modality treatment (Cummings, 1982; Cummings et al, 1984; Nigro, 1984). These metastases are becoming a more important cause of death as pelvic control rates improve. The toxicity of repeated courses of 5FU and MTC prevents its use as systemic adjuvant therapy and its efficacy is doubtful (Michaelson et al, 1983). Although distant metastases often respond to treatment with FUMIR, permanent control is achieved infrequently. The identification of effective systemic therapy, its possible interaction with programs such as FUMIR, and the selection of patients for systemic adjuvant therapy are presently unresolved problems.

Whether combined modality treatment is more effective than radical surgery or radical radiation is still uncertain. In comparison with the results of APR, of course, many patients preserve anal function when treated by FUMIR. In a recent report from the Memorial Hospital, Greenall et al (1985) found an absolute 5 year survival rate of 78 percent for 18 patients treated by combined therapy, and compared this to the 55 percent survival obtained in 103 patients treated previously at the same center by APR. However, in our study at PMH which also used historical controls, we failed to demonstrate any improvement in 5 year actuarial survival rates when FUMIR was compared to radical radiation therapy alone (Cummings et al, 1984). Several authors have reported good primary tumor control rates and survival, with low toxicity rates, with radiation alone (Cantril et al, 1983; Papillon, 1982), and have challenged the need for combined modality therapy such as FUMIR except for patients with advanced disease. Although staging systems and case selection undoubtedly differ from series to series, a rough comparison of treatment results according to the size of the primary anal tumor can be made. When the primary carcinoma was 5 cm or less in diameter, local control rates of 90 percent or better were reported with radical radiation alone (Cantril et al, 1983; Papillon, 1982), with radical radiation plus chemotherapy (Cummings et al, 1984), and with medium dose radiation plus chemotherapy and surgery (Nigro et al, 1984). For carcinomas larger than 5 cm in diameter, radical radiation plus chemotherapy produced local control in 90 percent (Cummings et al, 1984) compared to 68 percent following medium dose radiation and chemotherapy (Nigro, 1984) and 75 percent following radiation alone (Papillon, 1982). Whether these differences in local control or in survival rates are significant could only be resolved by formal comparative trials, and late toxicity rates following FUMIR also require longer term study.

Despite the several questions raised here, the widespread acceptance of combined modality therapy in its several forms represents a major advance in the management of anal canal carcinomas. Many patients are now being treated successfully with preservation of anal function, and the optimization of treatment programs can proceed in the context of improved patient comfort.

CARCINOMA OF THE ESOPHAGUS

The results of treatment of squamous cell carcinoma of the esophagus by surgical resection, radical radiation therapy, or resection plus adjuvant radiation alone have all been disappointing. Survival rates at 5 years are of the order of 5 percent to 10 percent (Earlam et al, 1980a, 1980b). Failure at the primary site caused persistent or recurrent dysphagia in as many as 80 percent of the patients we had treated by radical radiation (Beatty et al, 1979). Perioperative mortality rates averaging 29 percent detract from the results of surgical resection (Earlam et al, 1980a). The high complete tumor response rates to FUMIR observed in anal canal carcinoma led to the evaluation of this program in esophageal carcinoma. Again, these studies were empirical, and included both planned preoperative treatment, and concurrent chemotherapy and radical radiation.

The Detroit group elected to use preoperative treatment, and conducted a pilot study between 1977 and 1979 (Leichman et al, 1984). Thirty patients with localized and potentially curable esophageal cancer were treated by a protocol similar to that used by Nigro for anal cancer and shown in Figure 2. Twenty three patients underwent resection, and in six no residual cancer was found. Those with residual cancer received further chemotherapy and radiation postoperatively. Although the preoperative treatment was well tolerated, 7 of the 23 (30 percent) patients died following surgery of treatment related complications, thought to be due in part to the pulmonary effects of MTC (Steiger et al, 1981). Seven of the original 30 patients did not undergo surgery; one died suddenly prior to the planned operation, but only one of the other six developed local tumor recurrence. These authors noted that no patient in whom residual cancer was present after preoperative treatment survived longer than two years, with distant metastases being a major cause of death. Although the Eastern Cooperative Oncology Group has opened a randomized study to test the efficacy of preoperative FUMIR, the Detroit group has substituted cisplatin for MTC in their protocol, in an effort to identify chemotherapy more effective against distant metastases, and also to try to exploit the radiosensitizing potential of cisplatin (Douple, 1977). This revised protocol is now being evaluated in a randomized trial by the Southwest Oncology Group (SWOG) and Radiation Therapy Oncology Group (RTOG) (Leichman et al, 1984).

Our approach at PMH was to combine 5FU and MTC with our previously established radical radiation doses for esophageal carcinoma (Keane et al, 1985b), and others have used a similar strategy (Coia et al, 1984). In a pilot study we conducted between 1980 and 1983, 35 patients with biopsy proven squamous cell carcinoma without evidence of metastases outside the mediastinum were treated. All but one patient had a tumor which was either larger than 5 cm diameter, circumferential, or extending outside the esophagus. Because of the advanced age of many of the patients and concerns about possible toxicity, the physician could elect to treat by either uninterrupted radiation (Schedule B, Figure 1) or split course therapy (Schedule C, Figure 1). Fifteen patients were treated by a single course, and 20 by split course. The radiation fields encompassed the primary tumor with margins of about 5 cm above and below, without any attempt to cover the whole supero-inferior extent of the mediastinum. The spinal cord dose was restricted to 3500 to 4000 cGy. One patient treated by split course therapy died of aspiration pneumonia during treatment, and in six the chemotherapy was modified for the second course because of hematological toxicity. One patient died of late radiation pneumonitis. The majority of patients in this study were considered unsuitable for resection on medical grounds, or had surgically unresectable tumors. Only two patients underwent attempted resection for residual esophageal tumor following FUMIR, and both died postoperatively.

In half the patients (18/35) the tumor remained locally controlled. Each patient was matched with two controls from our earlier experience treated with radiation alone, and both the survival (2 year actuarial rate 30 percent versus 15 percent, $p = 0.004$) and local relapse-free rate (47 percent versus 20 percent, $p = 0.05$) were improved. The slope of the survival curve however was not changed, and the improvement may represent only a delay in death. Followup is still too short to determine whether the long term survival rate will be better with FUMIR. Continuous course therapy resulted in better survival at 2 years (48 percent versus 13 percent), and also in better local relapse-free rates (73 percent versus 29 percent), although the patient and tumor characteristics were not markedly different in the continuous and the split course groups.

Coia (1984) found that 6000 cGy in 6 to 7 weeks combined with two courses of 5FU and a single injection of MTC was well tolerated, and split course treatment was needed infrequently. At the time of their report, the median survival for 10 patients with Stage I or II disease was greater than 17 months, and only one patient had relapsed within the radiation field (followup 4 to 32 months).

In the study of preoperative FUMIR in which the radiation dose was 3000 cGy in 3 weeks, about 25 percent (6/23) had no residual carcinoma in the resected tissues (Leichman et al, 1984). An early report from the SWOG-RTOG trial with 5FU, cisplatin and 3000 cGy stated that 22 percent (19 of 86 patients) resection specimens were free of tumor (Leichman et al, 1984). In nonrandomized studies of radiation as single agent preoperative therapy, doses of 4500 cGy in 3 ½ weeks were associated with tumor-free resection specimens in 3 percent (3/101) (Marks et al, 1976), and 5000 to 6000 cGy in 5 to 7 weeks resulted in complete tumor clearance in 30 percent (7/23) but was associated with a high incidence of complications (Guernsey et al, 1969). Whilst direct comparison of these studies is not possible, there is at least a suggestion that combined radiation and chemotherapy may result in complete local tumor clearance with lower radiation doses. It is necessary to recall, however, that none of the studies of preoperative radiation alone produced strong evidence of improved survival.

As was the case with anal carcinoma, the one or two courses of 5FU and MTC do not appear to have reduced the risk of distant metastases. Metastases from esophageal carcinoma occur much more frequently than from anal carcinoma, and even if combined modality therapy does succeed in improving local control, survival rates will be influenced by whether effective systemic therapy can be developed.

The risk of distant metastases, the failure to improve cure rates when tumor was still present after preoperative FUMIR (Leichman et al, 1984), and the relatively high mortality rates following resection of residual esophageal cancer in both the Detroit and the Toronto series, suggest that patients in whom symptomatic residual tumor remains after treatment with FUMIR might be better to undergo palliative bypass rather than radical esophageal resection. Larger numbers of patients will need to be studied to determine the most appropriate treatment following chemotherapy and radiation, but an interim report from the SWOG-RTOG trial with 5FU, cisplatin and radiation reported death secondary to treatment in 8 of 86 patients (9.3 percent) who underwent surgery (Leichman et al, 1984).

Many centers have moved away from FUMIR as treatment for esophageal cancer. The substitution of cisplatin for MTC by some investigators has been noted above. Following the pilot FUMIR study described above, we found in a subsequent series of patients that 5FU plus uninterrupted radical radiation appeared to be just as effective as FUMIR (Keane, unpublished data), and Byfield (1980) has described the use of cyclical radiation and 5FU. At present, a multicenter study is being discussed in Canada in which 5FU and radical radiation would be compared with radical radiation alone.

CARCINOMA OF THE HEAD AND NECK

There have been numerous trials combining chemotherapy with radiation therapy in the management of head and neck cancer, but relatively few of these studies have included concurrent radiation and chemotherapy (Borgelt et al, 1978; Glick et al, 1981). Since local and regional failure after radiation therapy is a major cause of morbidity and mortality in advanced head and neck cancer, we were again influenced by the improved local control rates in anal cancer to test FUMIR against head and neck cancer.

Two pilot studies with FUMIR have been described. Kaplan et al (1985) treated patients with advanced stage tumors in several sites. The chemotherapy schedule was similar to that used at PMH, and was combined with 3000 cGy in 15 fractions in 3 weeks, followed by a two week rest period, after which the patients received a further 2000 cGy in 10 treatments without any chemotherapy. This schedule was well tolerated with severe mucositis, marrow depression or other complications occurring in only 10 to 15 percent of the 42 patients treated. In a subset of 31 patients treated with curative intent, complete clinical responses were recorded in 61 percent (19/31), and in 23 patients who underwent elective surgical resection the complete response rate judged by histological criteria was 52 percent (12/23). The authors did not report on the correlation between clinical and histological complete response. They did note that there was no apparent increase in surgical complications after this preoperative regimen.

At PMH, our previous experience with a policy of radical radiation therapy reserving surgical resection for the salvage of locally uncontrolled disease, had resulted in 5 year survival rates ranging between 0 and 50 percent in patients with advanced carcinoma of the larynx or hypopharynx, according to the extent of the tumor at the time of treatment (Harwood, 1982; Keane et al, 1982). In a preliminary study in patients with advanced carcinoma we found that uninterrupted treatment (Schedule B, Figure 1) produced intolerable mucositis and patients could not complete the 4 week course of radiation. In a pilot study in advanced laryngeal and hypopharyngeal carcinoma we therefore used a split course regimen (Schedule C, Figure 1) (Keane et al, 1985a). The radiation dose to the

Table 2. ADVANCED LARYNGEAL AND HYPOPHARYNGEAL CANCER

	Local Recurrence Free Rates	
	1 year	2 years
All patients	62%	51%
Larynx	56%	46%
Hypopharynx	85%	72%
	Regional Recurrence Free Rates	
All patients	74%	56%
Node negative	92%	92%
Node positive	61%	19%

spinal cord was restricted to 3500 to 4000 cGy. Ninety percent (51/57) of the patients had stage T_3 or T_4 tumors (U.I.C.C. staging), and 80 percent of the 33 patients with nodal metastases had nodes 3 cm or more in diameter. The toxicity of the split course program was acceptable, and 90 percent of the patients completed their radiation therapy, and 70 percent received two full courses of chemotherapy. The local and regional node recurrence free rates are shown in Table 2, and were considered similar to the results achieved with radiation therapy alone.

In 1982, in collaboration with several other centers, we undertook a randomized trial to compare split course FUMIR with radical radiation (Schedule A versus Schedule C, Figure 1) in patients with advanced laryngeal and hypopharyngeal carcinoma, and so far 180 of the 200 patients needed to complete this trial have been entered.

CARCINOMA OF THE CERVIX

Although radiation therapy has long been a major component of the treatment of squamous cell carcinoma of the cervix, patients who present with disease extending to both pelvic side walls, or to the rectum or bladder, have about a 75 percent risk of pelvic failure (Thomas, unpublished data). After it was observed in our series of patients with anal canal carcinoma that local control was maintained in 90 percent treated by FUMIR, a pilot study of this combination was commenced in patients with carcinoma of the cervix with the poor prognostic factors noted above (Thomas et al, 1984). When this study was opened in 1981, several laboratory studies examining the interactions of 5FU and radiation were available (Byfield et al, 1982) and it was decided that the treatment schedule should be designed to incorporate the premise that 5FU and radiation should be given concurrently, and that as much of the radiation as possible should be given during the period of drug infusion. The schedule chosen is shown in Figure 3. During the 4 day 5FU infusions with each part of the split course of radiation, the patients received two daily radiation fractions, each of 150 cGy. Following this the single daily fraction size was 180 cGy. The dose of MTC was reduced to 6 mg/m^2 because of concern about possible hematological toxicity. Early in the pilot study, patients received only 3 day infusions of 5FU, until the acute tolerance of the protocol had been established. Those patients with lymphangiographic evidence of metastases in the common iliac and/or the paraaortic nodes received paraaortic radiation, the fields otherwise being limited superiorly at the fifth lumbar vertebra. Where possible, FUMIR was followed by an intracavitary Cs-137 line source to deliver a dose of 4000 cGy at 2 cm from the applicator.

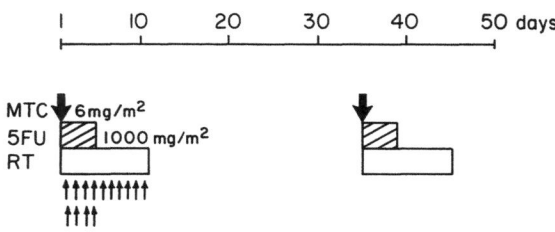

Figure 3. Treatment schedule for carcinoma of the cervix. MTC = Mitomycin C; 5FU = 5-Fluorouracil; RT = Radiation Therapy.

The preliminary results in the patients who have been followed for from 4 to 24 months showed that 74 percent (20/27) had an initial complete tumor response, but two later relapsed in the pelvis at 9 and at 12 months after therapy. The treatment was well tolerated, and only one patient developed a polymorph count less than 1000/mm^3, and seven patients developed moderate thrombocytopenia, although in no instance was treatment delayed by hematological depression. There were no cases of severe acute gastrointestinal toxicity, but three patients later developed severe sigmoid stricture or perforation (one fatal). It is of interest that the acute intestinal toxicity seems to have been less in the patients with cervix cancer than in those with anal cancer. Whether this is due to the differences in the treatment schedules, or to other factors, is not known.

The patients treated in this pilot study remain on followup. If the high rate of pelvic control is maintained, and the rate of late intestinal complications is not excessive, a randomized trial will be planned to compare combined chemotherapy and radiation with conventional radical radiation therapy.

CONCLUSION

The combination of 5FU, MTC and radiation is not a panacea for squamous cell carcinomas. The results from the treatment of these carcinomas in various tumor sites differ, as would be expected, since many factors other than the histological appearance of a carcinoma determine its natural history. The principle intent of the use of this combination in current practice appears to be to improve control in the primary tumor and regional nodes within the irradiated volume, through an interaction between the chemotherapy agents and the radiation. Although some schedules call for a second course of 5FU which is not concurrent with the radiation, it seems unrealistic to expect only two courses of treatment with drugs which have demonstrated limited efficacy as single agent therapy to have any marked effect against occult distant metastases, and there is still a need to develop effective systemic therapy for all of the tumor sites discussed.

The randomized trials already opened for some tumor sites will help to determine the role of FUMIR. In other sites, such as anal carcinoma, the very good local response rates allowing preservation of anal function, and the suggestion that survival rates with FUMIR are no worse than with radical radiation or radical surgery alone, will probably be accepted by most physicians without the need for formal randomized trials.

It should be clear from this overview of the use of FUMIR that many questions about this combination are still unanswered. There are many reports of improved response rates, but these may not necessarily be followed by improvements in survival, and more information is also needed on late toxicity. Much of the present empirical adjustment of treatment schedules in clinical practice could be better directed if reliable experimental models of these tumor systems could be developed, so that the interactions of chemotherapy agents and radiation could be better understood.

ACKNOWLEDGEMENTS

The authors gratefully acknowledge the assistance of Miss E. Eisenreich in the preparation of this manuscript.

REFERENCES

Beatty, J.D., DeBoer, G., and Rider, W.D., 1979, Carcinoma of the esophagus: Pretreatment assessment, correlation of radiation treatment parameters with survival and identification and management of radiation treatment failure, Cancer, 43:2284.

Borgelt, B.B., and Davis, L.W., 1978, Combination chemotherapy and irradiation for head and neck cancer: a review, Cancer Clin Trials, 1:49.

Byfield, J.E., Barone, R.M., Mendelsohn, J., Frankel, S., Quinol, L., Sharp, T., and Seagren, S., 1980, Infusional 5-fluorouracil and x-ray therapy for non-resectable esophageal cancer, Cancer, 45:703.

Byfield, J.E., Barone, R.M., Sharp, T.R., and Frankel, S.S., 1983, Conservative management without alkylating agents of squamous cell anal cancer using cyclical 5FU alone and x-ray therapy, Cancer Treat Rep, 67:709.

Byfield, J.E., Calabro-Jones, P., Klisak, I., and Kulhanian, F., 1982, Pharmacologic requirements for obtaining sensitization of human tumor cells in vitro to combined 5-fluorouracil or Ftorafur and x-rays, Int J Radiat Oncol Biol Phys, 8:1923.

Cantril, S.T., Green, J.P., Schall, G.L., and Schaupp, W.C., 1983, Primary radiation therapy in the treatment of anal carcinoma, Int J Radiat Oncol Biol Phys, 9:1271.

Coia, L.R., Engstrom, P.F., Paul, A., Gallagher, M.J., Stoll, D., Catalano, R., and Richter, M.P., 1984, A pilot study of combined radiotherapy and chemotherapy for esophageal carcinoma, Am J Clin Oncol, 7:653.

Cummings, B.J., 1982, The place of radiation therapy in the treatment of carcinoma of the anal canal, Cancer Treat Rev, 9:125.

Cummings, B.J., Harwood, A.R., Keane, T.J., Thomas, G.M., and Rider, W.D., 1980, Combined treatment of squamous cell carcinoma of the anal canal: Radical radiation therapy with 5-Fluorouracil and mitomycin C, a preliminary report, Dis Colon Rectum, 23:389.

Cummings, B., Keane, T., Thomas, G., Harwood, A., and Rider, W., 1984, Results and toxicity of the treatment of anal canal carcinoma by radiation therapy or radiation therapy and chemotherapy, Cancer, 54:2062.

Douple, E.B., 1977, Therapeutic potentiation in a mouse mammary tumor and an intracerebral rat brain tumor by combined treatment with cis-dichloroplatinum II and radiation, J Clin Hemat Oncol, 7:585.

Earlam, R., and Cunha-Melo, J.R., 1980a, Oesophageal squamous cell carcinoma. I. A critical review of surgery, Br J Surg, 67:381.

Earlam, R., and Cunha-Melo, J.R., 1980b, Oesophageal squamous cell carcinoma. II. A critical review of radiotherapy, Br J Surg, 67:457.

Fu, K.K., 1985, Biological basis for the interaction of chemotherapeutic agents and radiation therapy, Cancer, 55:2123.

Glick, J.H., and Taylor, S.G., 1981, Integration of chemotherapy into a combined modality treatment plan for head and neck cancer: a review, Int J Radiat Oncol Biol Phys, 7:229.

Gollin, F.F., Ansfield, F.J., Brandenburg, J.H., Ramirez, G., and Vermund, H., 1972, Combined therapy in advanced head and neck cancer, Am J Roentgenol, 114:83.

Greenall, M.J., Quan, S.H.Q., Urmacher, C., and DeCosse, J.J., 1985, Treatment of epidermoid cancer of the anal canal, Surg Gynecol Obstet (in press).

Guernsey, J.M., Doggett, R.L.S., and Mason, G.R., 1969, Combined treatment of cancer of the esophagus, Am J Surg, 117:157.

Harwood, A.R., 1982, Cancer of the larynx - the Toronto experience, J Otolaryngol, 11:11.

Kaplan, M.J., Hahn, S.S., Johns, M.E., Stewart, F.M., Constable, W.C., and Cantrell, R.W., 1985, Mitomycin and fluorouracil with concomitant radiotherapy in head and neck cancer, Arch Otolaryngol, 111:220.

Keane, T.J., Harwood, A.R., Beale, F.A., Cummings, B.J., Payne, D.G., and Rawlinson, E., 1985a, A pilot study of mitomycin C and 5-Fluorouracil infusion combined with split course radiation therapy for advanced carcinomas of the larynx and hypopharynx, (submitted for publication).

Keane, T.J., Harwood, A.R., Elhakim, T., Rider, W.D., Cummings, B.J., Ginsberg, R.J., and Cooper, J.C., 1985b, Radical radiation therapy with 5-Fluorouracil infusion and mitomycin C for oesophageal squamous carcinoma, Radiother Oncol, (in press).

Keane, T.J., Hawkins, N.V., Beale, F.A., Cummings, B.J., Harwood, A.R., Payne, D.G., and Rider, W.D., 1982, Carcinoma of the hypopharynx. Results of primary radical radiation therapy, Int J Radiat Oncol Biol Phys, 9:659.

Leichman, L., Steiger, Z., Seydel, H.G., and Vaitkevicius, V.K., 1984, Combined preoperative chemotherapy and radiation therapy for cancer of the esophagus: the Wayne State University, Southwest Oncology Group and Radiation Therapy Oncology Group experience, Sem Oncol, 11:178.

Marks, R.D., Schruggs, H.J., and Wallace, K.M., 1976, Preoperative radiation therapy for carcinoma of the esophagus, Cancer, 38:84.

Michaelson, R.A., Magill, G.B., Quan, S.H.Q., Leaming, R.H., Nikrui, M., and Stearns, M.W., 1983, Preoperative chemotherapy and radiation therapy in the management of anal epidermoid carcinoma, Cancer, 51:390.

Nigro, N.D., 1984, An evaluation of combined therapy for squamous cell cancer of the anal canal, Dis Colon Rectum, 27:763.

Nigro, N.D., Vaitkevicius, V.K., and Considine, B., 1974, Combined therapy for cancer of the anal canal: a preliminary report, Dis Colon Rectum, 17:354.

Papillon, J., 1982, "Rectal and Anal Cancers", Springer-Verlag, Berlin.

Peckham, M.J., and Collis, C.H., 1981, Clinical objectives and normal tissue responses in combined chemotherapy and radiotherapy, Bull Cancer (Paris), 68:132.

Reich, S.D., 1979, Clinical pharmacology of mitomycin C, in "Mitomycin C. Current Status and New Developments," S.K. Carter and S.T. Crooke, ed., Academic Press, New York.

Rockwell, S., 1982, Cytotoxicities of mitomycin C and x-rays to aerobic and hypoxic cells in vitro, Int J Radiat Oncol Biol Phys, 8:1035.

Rosenberg, J.C., Schwade, J.G., and Vaitkevicius, V.K., 1982, Cancer of the esophagus, in "Cancer, Principles and Practice of Oncology," V.T. Devita, S. Hellman, and S.A. Rosenberg, eds., Lippincott, Philadelphia.

Seifert, P., Baker, L.H., Reed, M.L., and Vaitkevicius, V.K., 1975, Comparison of continuously infused 5-Fluorouracil with bolus injection in treatment of patients with colorectal adenocarcinoma, Cancer, 36:123.

Sischy, B., Remington, J.H., Hinson, E.J., Sobel, S.H., and Woll, J.E., 1982, Definitive treatment of anal canal carcinoma by means of radiation therapy and chemotherapy, Dis Colon Rectum, 25:685.

Steiger, Z., Franklin, R., and Wilson, R.F., 1981, Eradication and palliation of squamous cell carcinoma of the esophagus with chemotherapy, radiotherapy, and surgical therapy, J Thorac Cardiovasc Surg, 82:713.

Teicher, B.A., Lazo, J.S., and Sartorelli, A.C., 1981, Classification of antineoplastic agents by their selective toxicities toward oxygenated and hypoxic tumor cells, Cancer Res, 41:73.

Thomas, G., Dembo, A., Beale, F., Bean, H., Bush, R., Herman, J., Pringle, J., Rawlings, G., Sturgeon, J., Fine, S., and Black, B., 1984, Concurrent radiation, mitomycin C and 5-Fluorouracil in poor prognosis carcinoma of cervix: preliminary results of a phase I-II study, Int J Radiat Oncol Biol Phys, 10:1785.

TREATMENT OF BLADDER CARCINOMA WITH CONCOMITANT INFUSION

CHEMOTHERAPY AND IRRADIATION

M. Rotman[1], R. Macchia[2], M. Silverstein[1], K. Choi[1],
J. Rosenthal[3], A. Braverman[3], and H. Aziz[1]

SUNY Downstate Medical Center, [1]Dept. of Radiation Oncology
[2]Urology, [3]Medicine, 450 Clarkson Ave., Brooklyn, New York

INTRODUCTION

In 1983 carcinoma of the urinary bladder was responsible for over 10,000 deaths in the United States[1]. Despite advances in treatment and detection of the disease, this death rate has remained virtually unchanged for the last 30 years. Transurethral resection, partial cystectomy and interstitial implant are appropriate treatments for low grade, low stage tumors, while pre-operative irradiation followed by planned cystectomy is the accepted modality of treatment for high stage, deeply invasive carcinomas. Patients who demonstrate complete response in the surgical cystectomy specimen carry a better prognosis[2,3,4,5]. Blandy in 1980, showed that patients achieving complete response within 6 months of curative irradiation had a 5 year survival rate of 56%[2]. Therefore, if the complete response rate can be improved then hopefully this will translate into prolonged survival.

Cystectomy leaves the patient with a permanent ileal conduit and leaves the male impotent. Alternative methods of treatment have been sought that would enable the patient to maintain normal bladder function while still providing a good chance of long term disease free survival, with minimal morbidity.

In recent years, chemotherapeutic agents used as radiation sensitizers have improved tumor clearance in certain epithelial malignancies[6,7,8,9]. Byfield using 5-FU as a radiosensitizer has shown improved complete response rate in the treatment of esophageal carcinoma[8], while Shipley using Cis-platinum as an IV bolus achieved a CR rate of 76% in the treatment of bladder carcinoma[10].

In an attempt to increase tumor clearance and bladder preservation without increased morbidity, a pilot study using irradiation and concomitant continuous 5-FU infusion, and Mitomycin C as IV bolus in the treatment of invasive high grade bladder carcinomas was started at Downstate Medical Center. This report shows our preliminary results regarding tumor response, survival and morbidity of treatment.

MATERIALS AND METHODS

Between July 1980 and December 1984 a total of 16 patients, 12 male and 4 female, between the ages of 49 and 95 years (median age 68 years) were included in the study. Four patients had stage C disease, six had Stage B_2, five had Stage B_1 and one had Stage A disease. Of 14 patients who had transitional carcinoma, two had Grade 2 and twelve had Grade 3-4. The other two patients had Grade 3 squamous cell carcinoma. Six of sixteen patients had more than 50% of the bladder involved with tumor, seven had 50% and three less than 25%. Nine out of sixteen patients had hydronephrosis on IVP.

5-FU was given as a continuous 96 hour infusion (25 mg/kg/day) to all patients during weeks 1,4,7 or 2,5, and 8 of the treatment. In addition, Mitomycin C was delivered to four patients as an IV bolus (10 mg/m^2) on day one of radiation therapy. The total radiation dose was 60-65 Gy in 33-35 fractions over 6 1/2 to 7 weeks. All patients were followed by serial cystoscopy, biopsy and urine cytology.

RESULTS

16 patients were analyzed for tumor response and complications. 13/16 (81%) had a complete response (CR) manifested by total visual clearance of tumor and negative bladder biopsy within 6 months post-treatment. An additional two patients became free of disease at 11 and 17 months, for a total clearance rate of 94% (15/16 pts.). 14/15 patients who achieved CR maintained local control. The one patient who had a complete response but did not maintain local control later showed diffuse carcinoma in situ and a single microscopic focus of invasive tumor in the bladder and a transitional cell tumor in the ureter. The one patient who was a non-responder had a histology of squamous cell carcinoma. Both patients received a salvage cystectomy and are alive and well at 20 months and 17 months respectively without evidence of disease. 10/15 patients who achieved total tumor clearance and have been followed for between 8 and 50 months have maintained negative urine cytologies. Of the 3 patients who still have positive cytology, two have developed a 2nd tumor in the upper collecting system and the third had extensive CIS. Two patients have not had any follow-up cytological examination. One patient was excluded from the study because of the development of distant metastases prior to completing therapy.

11/15 patients (73%) are alive 8 months to 50 months after initial treatment with a median survival of 22 months. 8/11 patients are alive without evidence of disease. Two patients who had salvage cystectomy for persistant disease are also alive and well without evidence of disease at 17 and 20 months. One patient is alive with distant metastasis. 9/11 patients who are alive have retained a functional tumor-free bladder. Of the four patients who have died, one developed a 2nd primary in the colon and died of metastatic colon carcinoma at 50 months. He maintained a functional tumor-free bladder. The other three patients were all free of tumor in the bladder, had negative cytologies and died of distant metastatic disease between 8-23 months with a median survival of 9 months.

The treatment regimen was well-tolerated with mild to moderate anorexia, nausea, urinary frequency and diarrhea during the treatment course. These mild to moderate complications during the period of treatment were well controlled with usual medication. Four patients had developed late complications, which included contracted bladder in two patients and hemorrhagic cystitis in two patients.

DISCUSSION

Potentiation of tumor cell killing has been sought from the combination of irradiation and a group of chemotherapeutic agents classified as radiosensitizers. The resultant antineoplastic effect has been described as additive or synergistic[11,12]. The latter is defined as cell killing at a level greater than predicted by the product of the cytotoxicity exerted by each modality administered alone[11].

Considerable laboratory and clinical experience has been accumulated utilizing 5-fluorouracil as a radiosensitizer. Heidelberger, et al in 1958 showed that ineffective doses of either irradiation or 5-FU could be made curative for certain rodent tumors if both treatments were combined[13]. Vermund, et al[14] in 1964 confirmed Heidelberger's results with in-vitro experimentation. The reason for enhanced cell killing with a combination of irradiation and 5-FU is not well understood. Vietti, et al[15] performing in-vivo experiments on the rodent AKR system, hypothesized that the increased cell killing was due to inhibition of repair of sub-lethal x-ray damage. This later was refuted by Byfield utilizing in-vitro experimentation with HT.29 cells[11]. Byfield's data also suggested that for synergism to occur between the effects of 5-FU and irradiation, the cells should be continuously exposed to 5-FU at least 48 hours after irradiation. This investigation also noted that sensitization can be achieved only by infusion; bolus injection resulted in only additive effects. Seifert[16], on the basis of a prospective randomized trial established that the response rate for 5-FU delivered as an infusion was better over bolus injection. Furthermore, the tolerance of the cumulative doses and toxicity was improved as a result of infusional therapy.

Mitomycin C has been used as a single cytotoxic agent in the treatment of advanced bladder cancer and has been reported by Early, et al[17] to induce a 13% response rate in patients with metastatic disease. It has been used as an intravesical agent in the treatment of recurrent or persistant superficial bladder cancer as well as primary therapy for carcinoma in-situ[18]. Mitomycin C delivered as an IV bolus has been combined with irradiation and continuous 5-FU in the treatment of squamous cell carcinoma of the esophagus and anal canal. Complete eradication of tumor in the resected esophagus has been found in 83% of patients[8] and in over 80% of patients treated for cancer of the anal canal[7].

Utilizing the above method of combined modality therapy a 94% complete response rate was obtained in our pilot study of 16 patients. Bladder presevation has been possible in 9 of 11 patients (82%) alive with or without disease.

Blandy et al[2] has postulated that those patients who show complete regression of tumor within 6 months of full course irradiation should be followed conservatively. Only those patients who failed to achieve a complete response or developed a tumor recurrence after 6 months of treatment should be considered for salvage cystectomy. Two patients in our series underwent salvage cystectomy for persistent disease and are presently alive and well. Operative mortality from salvage cystectomy has been reported by Blandy, et al[2] and Goodman, et al[19] to be 11% which is only slightly higher than the 8% operative mortality reported for pre-operative irradiation and radical cystectomy.

The acute complications from the combined treatment regimen were mild to moderate and consisted of anorexia, nausea and diarrhea and were suprisingly controlled considering the advanced ages of many of our patients. Late complications consisted of two cases of hemorrhagic cystitis and two patients with contracted bladder.

SUMMARY

Patients who exhibit complete response following irradiation of bladder carcinoma have a better survival rate[2,3,4,5]. In an effort to achieve this goal at Downstate Medical Center, 5-fluorouracil (5-FU) has been utilized as a radiosensitizer. 16 patients who were treated with 60-65 Gy and concomitant continuous 5-FU infusion with or without Mitomycin C IV bolus achieved a complete response rate of 94%. Seventy three percent of patients are alive at 22 months and have retained a functional, tumor-free bladder. Salvage cystectomy was performed on two patients for persistant disease. There were two cases of hemorrhagic cystitis and two patients have a contracted bladder. It it hoped that this impressive complete response rate will translate into increased long term survival while retaining a functional, tumor-free bladder.

REFERENCES

1. Silverberg E: Cancer Statistics, 1983. CA Jan-Feb 33:9-25, 1983.
2. Blandy JP, England HR, Evans SJ, Hope-Stone HF, Mair GM, Mantel BS, Oliver RT, Paris AM, Risdon RA: T3 Bladder Cancer - The Case for Salvage Cystectomy. Br J Urol 52:506-510, 1980.
3. Bloom, HJG, Henry W, Wallace D, Skeet R: Treatment of T3 Bladder Cancer: Controlled Trial of Pre-Operative Radiotherapy and Radical Cystectomy vs Radical Radiotherapy. Second Report and Review Brit J Urol 54:136-151, 1982.
4. van der Werf-Messing G, Friedell GH, Menon RS, Hop WCJ, Wassif SB: Carcinoma of the Urinary Bladder T3NXMO Treated by Pre-operative Irradiation Followed by Cystectomy. Int J Radiat Biol Phys 8:1849-1855, 1982.
5. Batata MA, Chu FCH, Hilaris BS, Kim YS, Lee MZ, Chung S, Whitmore WF: Factors of Prognostic and Therapeutic Significance in Patients with Bladder Cancer. Int J Radiat Oncol Biol Phys 7:575-579, 1981.
6. Buroker T, Nirgo N, Considine B, Vaitkevicius VK: Mitomycin-C, 5-fluorouracil, and Radiation Therapy in Squamous (Epidermoid) Cell Carcinoma of Anal Canal. In Mitomycin-C: Current Status and New Developments, SK Carter and ST Crooke Eds, New York, Academic Press, 1979.
7. Sischy B, Remington JH, Hinson EJ, et al: Definitive Treatment of Anal Canal Carcinoma by Means of Radiation Therapy and Chemotherapy. Disease of Colon and Rectum, 7:685-688, 1982.
8. Byfield JE, Barone R, Mendelsohn J, Frankel S, Quinol L, Sharp T, Seagren S: Infusion 5-FU and X-ray Therapy for Nonresectable Esophageal Cancer. Cancer 45:703-708, 1980.
9. Moertel CG, Childs DS, Jr., Reitemeier RJ, et al: Combined 5-fluorouracil and Supervoltage Radiation Therapy of Locally Unresectable Gastrointestinal Cancer. Lancet 2:865-867, 1969.
10. Shipley WU, Coombs LJ, Einstein AB, Socoway MS, Wassman Z, Prout George RJR and National Bladder Cancer Collaborative Group A: Cisplatin and Full Dose Irradiation for Patients with Invasive Bladder Carcinoma: A Preliminary Report of Tolerance and Local Response. The J of Urol 132:899-903, 1984.
11. Byfield JE, Calabro Jones P, Klisak I, Kulhanian F: Pharmacologic Requirements for Obtaining Sensitization of Human Tumor Cells in vitro to Combined 5-fluorouracil or Ftorafar and X-rays. Int J Radiat Oncol Biol Phys 8:1923-1933, 1982.
12. Phillips TL, and Fu KK: Quantification of Combined Radiation Therapy and Chemotherapy Effects on Critical Normal Tissue Cancer 37:1186-1200, 1976.
13. Heidelberger C, Greisbach L, Montag BJ, Mooren D, Cruz D, Schnitzer RJ, Grunberg E: Studies in Fluorinated Pyrimidin II. Effects of Transplanted Tumor Ca Res. 18:305-317, 1958.
14. Vermund H, Hodgett J, Ansfield FJ: Effects of Combined Chemotherapy on Transplanted Tumor in Mice. Amer J Roentgen 85:559-567, 1961.
15. Vietti J., Eggerding F., Valeriote F: Combined Effect of Radiation

and 5-fluorouracil on Survival of Transplanted Leukemia Cells. *J Natl Ca Inst* 47:865-870, 1971.
16. Seifert P, Baker LH, Reed ML: Comparison of Continuous Infused 5-FU with Bolus Injection in Treatment of Patients with Colorectal Adenocarcinoma. *Cancer* 36:123-128, 1975.
17. Early K, Elias EG, Mittleman A. et al: Mitomycin C in the Treatment of Metastatic Transitional Cell Carcinoma of the Bladder. *Cancer* 31:1150-1153, 1973.
18. Soloway MS: Surgery and Intravesical Chemotherapy in the Management of Superficial Bladder Cancer. *Seminars in Urol* 1:23-33, 1983.
19. Goodman GB, Hislop TG, Elwood JM, Balfour J: Conservation of Bladder Function in Patients with Invasive Bladder Cancer Treated by Definitive Irradiation and Selective Cystectomy. *Int J Radiat Oncol Biol Phys* 7:569-573, 1982.

A UROLOGIST'S VIEWPOINT: TREATMENT OF INVASIVE BLADDER

CANCER BY THE XRT/5FU PROTOCOL

      Richard J. Macchia, and
      Gobind Laungani

      S.U.N.Y. - Downstate Medical Center
      Department of Urology - Box 79
      Brooklyn, N.Y. 11203

    Primary transitional cell carcinoma (TCC) localized to the urinary bladder occurs as: A, carcinoma in situ (CIS), B, superficial disease penetrating the basement membrane but not invasive of detrusor muscle, C, disease invasive of the detrusor muscle, or a mixture of these. CIS is most frequently treated with intravesical chemo or immunotherapy. Radical cystectomy is utilized for treatment failures. Superficial disease is treated by transurethral resection (TUR-BT) followed by intravesical chemo or immunotherapy.

    The current most widely advocated therapy in the USA for invasive disease is total cystectomy usually preceded by adjuvant external radiation therapy (XRT). The procedure involves removal of the bladder, prostate, seminal vesicles, distal ureters, and frequently, the urethra. Urinary diversion is usually accomplished by ileal conduit. Cystectomy renders males impotent. In women an abdominal hysterectomy and a partial anterior vaginectomy is required. The mortality rate ranges from 1-14%. A significant rate of serious early and late complications exists. Despite the formidable nature of the procedure 50% of patients die of metastatic disease. The roles of pelvic lymphadenectomy and preoperative radiation remain controversial.

    Approximately 40,000 new cases of this disease are diagnosed in the United States each year. Approximately 80% of patients have disease localized to the bladder at the time of the diagnosis. This is in sharp contrast to the situation in prostate cancer where only about 20% of the 80,000 newly diagnosed cases are localized. Therefore, the ability to control localized disease would lead to a significant improvement in overall survival. Bladder cancer tends to recur locally in a high percentage of the patients regardless of the initial therapy. A variety of factors have been identified which places a patient in the high risk category, for example: maleness, multiple tumors either at the same time or sequentially, size of tumor, location of tumor within the bladder, grade of tumor, concurrent carcinoma in situ. The recurrence of the carcinoma can be anywhere in the urothelial system. The urothelium is the lining of the luminous portion of the urinary tract from the tip of the calyces to the urethral meatus. Recurrence can be at the site of the original lesion and/or at any other site in the urothelium. Most recurrences of bladder carcinoma are within the bladder.

In our protocol XRT/5FU was utilized only in those patients who refused cystectomy or who were not candidates for cystectomy because of their medical condition. The problem facing the urologist is a complete elucidation of the patient's tumor status. This is especially important in a prospective study. The amount of information gathered is voluminous and precludes inclusion in this brief article. However, of obvious importance is a definitive description of the size and location of any visible lesions. The histologic nature of the lesion together with documentation of the depth of invasion is critical. These can be determined only by resection of the tumor and the subjacent bladder (detrusor) muscle. In addition, so-called bladder mapping was undertaken to determine whether other areas of the bladder contained carcinoma in situ. This was done by random or selected site cold-cup mucosal biopsies. The percentage of bladder surface involved by visible tumor was estimated. The diagnostic procedure of the TUR-BT is frequently also the therapeutic measure. However, in some cases, the surface area of the tumor precluded complete resection. The tumor status of the kidneys, ureters, and urethra was carefully evaluated. The patient was then placed in the appropriate clinical stage.

Excretory urography, cystoscopy, cytology and computerized tomography were the main imaging methods but specialized staging and follow-up techniques were needed in some patients. For example, retrograde ureteropyeloscopy allows direct visualization and biopsy of the ureters and the renal pelvis. Selective ureteral cytology or ureteral brush was routinely utilized. Transrectal, transperineal or transurethral biopsy of the prostate was occasionally required as concomitant transitional cell or adenocarcinoma of the prostate is well known. For certain patients percutaneous antegrade techniques were required. Percutaneous techniques involve the passage of a tube through the flank into the kidney. This opening is then enlarged until appropriate instrumentation can be introduced. Some patients required percutaneous aspiration cytology of pelvic or abdominal lymph nodes. No patient was subjected to pelvic or abdominal lymphadenectomy as a purely staging procedure.

The next problem is appropriate follow-up after XTR/5FU. The initial evaluation was performed at one, two, or three months after the completion of radiation therapy. Thereafter, follow-up was performed at regular three month intervals. Follow-up consisted in all cases of physical examination, chest x-ray, cystoscopy and urinary cytology. These modalities were supplemented where indicated by the above techniques. All cystoscopies were performed under general or regional anesthesia.

The necessity for the repetitive invasive procedures to follow these patients is one of the drawbacks to this protocol. Repetitive instrumentation together with the XRT/5FU treatment and concurrent associated incidental disease, such as BPH and urethral stricture, may render the lower urinary tract progressively more difficult to instrument and endanger the continence mechanism. The bladder mucosa after XRT/5FU is rarely similar to the pristine state. Erythematous areas usually abound. A decision must be made as to which of these areas to subject to a mucosal biopsy. In addition to the possible risk of disseminating the disease, multiple biopsy irritates the bladder and contributes to the development of lower tract irritability.

One must be prepared to deal with the inevitable complications: periodic gross microscopic hematuria, lower GU and GI tract irritability, decreased libido and impotency. An objective study of the quality of life for patients who are rendered free of disease by the use of XRT has yet to be published. For those patients who fail this protocol, "salvage" cystectomy is undertaken where feasible. After a full course of radiation therapy and adjuvant chemotherapy the procedure is rendered more tedious and the potential for complication is greater.

Our goal here was to attempt to improve the survival for those patients who count not undergo the current treatment of choice. The results as outlined in our accompanying article indicate that this technique is promising, but problems abound and questions persist. It appears that XRT/5FU has little if any effect on CIS, therefore, intravesical chemo and immunotherapy are important adjuvants. Our work cannot answer the questions as to whether or not there was true radiosensitization. The phenomena of radioresistance, radiosensitization and radioprotection are being intensively investigated in the laboratory and clinically in several centers. We must persist in developing organ preserving therapy.

REFERENCES

1. Radwin, H.M. Radiotherapy and bladder cancer: a critical review. J. Urol. 124:43-46.
2. Rotman, M., Macchia, R., Silverstein, M., et al. Treatment of bladder carcinoma with concomitant infusion chemotherapy and irradiation - in this volume.

CONCOMITANT RADIATION THERAPY AND DOXORUBICIN BY CONTINUOUS INFUSION IN ADVANCED MALIGNANCIES - A PHASE I-II STUDY - EVIDENCE OF SYNERGISTIC EFFECT IN SOFT TISSUE SARCOMAS AND HEPATOMAS

C. Julian Rosenthal, Marvin Rotman and Inder Bhutiani

Division of Medical Oncology and Department of Radiation Oncology, Downstate Medical Center, State University of New York, Brooklyn, New York

INTRODUCTION

The current lack of effective therapy against most malignant tumors in advanced stages is due, primarily, to its lack of specific cytotoxicity for neoplastic cells.

In order to circumvent this problem various chemotherapeutic agents have been used in combinations aiming to increase their overall therapeutic index.

Potentiation of antineoplastic effects has also been sought from the combined administration of chemotherapy agents and radiation therapy (1). The benefit of radiation enhancement by chemotherapy agents have been described as additive or synergistic (2). The latter refers to a situation in which the effect of the combined treatment is greater than the sum of the cell kill expected from the two modalities administered alone. The term dose effect factor (DEF), defined as the ratio of radiation dose required in the absence of a drug to the radiation dose required to cause the same level of damage in its presence was introduced to assess the sensitizing effect of various agents (2,3). Among all antineoplastic agents, the antibiotics had the higher DEF for both neoplastic cells and normal tissues; consequently, their combination with radiation to large areas or vital structures could be lethal. However, preliminary clinical studies (4,5) showed that single pulses of adriamycin in conventional doses (40-60 mg/m^2 q 3 wks.) or in lower doses (12 mg/m^2) administered concurrently with radiation therapy were well tolerated.

Recent in vitro studies (6) showed that increased exposure of tumor cells to adriamycin yielded equal cytotoxic effect at one thousand the usual dose. Further, adriamycin administered by continuous intravenous infusion for up to 96 hours at a dose of 20 mg/m^2/24 hours was found to have significantly less cardiotoxic effects (7). Based on these data, we postulated that concomitant administration of low dose adriamycin in continuous infusion with daily pulses of radiotherapy to metastatic, recurrent or unresectable malignant lesions could improve the therapeutic index of these two modalities of treatment.

At the time this study was started there were no clinical data concerning the use of adriamycin in continuous infusion nor any reported clinical experience with the concomitant use of this drug in infusion and radiation therapy.

For this reason this investigation was initially developed as a phase I study with three objectives: 1) to establish the dose and the duration of administration of doxorubicin infusion having the best therapeutic index; 2) to determine the dose of radiation which in combination with the optimal dose of doxorubicin administered by continuous infusion would have the best therapeutic index; 3) to establish the nature and the incidence of side effects of these combined modalities of therapy. The therapeutic effect of the optimal doxorubicin infusion dose in combination with concomitant radiotherapy was then assessed in a preliminary phase II study; it was found to have a significant enhancing effect in a few cases of hepatomas and in most of the cases of recurrent and metastatic soft tissue sarcomas.

MATERIALS AND METHODS

1. Patient population

Thirty two patients were entered in this study over a period of 30 months; thirteeen received doxorubicin infusion alone, while another nineteen received doxorubicin infusion with concomitant radiation to some of their measurable lesions. Active infections or sepsis, bone marrow suppression, cardiac arrythmias or failure were contraindications for entering patients in this study. There were 14 females and 18 males varying in ages between 29 and 75 years old. All patients had advanced metastatic and/or recurrent loco-regional disease measurable by palpation, x-ray or CT scanning.

The diagnosis of patients who received continuous infusion of doxorubicin alone included the following malignancies: gastric carcinoma (3 cases) infiltrating ductal carcinoma of the breast (1), bronchogenic adenocarcinoma (3), soft tissue sarcoma (3), transitional cell carcinoma of the bladder (2) and undifferentiated carcinoma (1). The concomitant combined modality therapy was administered to the following groups of patients: seven with adenocarcinoma including carcinoma of the stomach (2), colon (2) and breast (3), four with hepatocellular carcinoma, one with transitional cell carcinoma of the bladder and seven with soft tissue sarcomas. The characteristic, clinical and histologic features of these 19 patients are listed on table I (see appendix).

All patients had advanced metastatic or recurrent disease with invasion of regional lymphatics or of neighboring structures. Among patients with soft tissue sarcomas two of the patients had grade II (G_2) moderately differentiated tumors (both with leiomyosarcoma) while the other five had G_3, poorly differentiated tumors as defined by the American Joint Committee Task Force on soft tissue sarcomas classification (8).

The performance status of these patients at their accession in this study varied between 0 and 3 on the ECOG scale.

All but two of these patients failed to respond to chemotherapy prior to their entry in this study. Doxorubicin was part of their early treatment in all but one of the cases; it was administered monthly as an intravenous bolus injection, alone or in combination with other drugs (table I) up to a total dose varying between 280 and 460 mg/m^2. Four patients have also received radiation therapy at least three months prior to their entry in this study; it was administered to the measurable lesions, treated in this study, with only a transient effect (in patient OV) or no effect at all (to patient ME) on the growth of the lesion. All patients signed appropriate informed consent.

2. Therapeutic regimens

Doxorubicin was delivered through silicone elastomer catheters (Deseret-Parke Davis, Sandy, Utah) inserted in the superior vena cava via the external jugular vein for each of the 5 day cycles of administration or through a Hickman's catheter (10) inserted in the subclavian vein. The total daily dose of doxorubicin was mixed in 1000 ml of 5% dextrose solution in half normal saline and delivered over 24 hours at a controlled rate by an IVAC pump (IVAC-Sacramento, Ca.) or was delivered slowly by a battery activated pump (Cormed-Medina, NY) attached to the Hickman's catheter.

An initial dose of 9 mg/m^2/d of doxorubicin, found to be appropriate in prior studies (9) on the intra-arterial administration of doxorubicin, was given by continuous 5 day infusion to three patients with measurable metastatic lesions, originated from a gastric adenocarcinoma, an adenocarcinoma of the lung and a liposarcoma of the right arm, respectively. The infusion was repeated every 3-4 weeks as soon as patient's WBC and platelet count were above 80% of their initial normal values. The dose was escalated in increments of 2 mg/m^2 every cycle until unwanted side effects developed.

Ten more patients with advanced measurable malignant tumors, have then received the dose of doxorubicin which was found to have optimal therapeutic index defined as the dose inducing best results accompanied by non-life threatening side effects. Progression of measurable disease or development of a decrease of the ventricular ejection fraction were considered the end point of this study for each evaluable patient.

In the second phase of this study radiation therapy was added to doxorubicin administered by continuous 5 day infusion at a dose of 12 mg/m^2/d, slightly lower than that found to be optimal when administered alone. Radiation was initially delivered in 5 daily fractions of 100 rads per session and repeated each time doxorubicin was administered (q 3-4 weeks). The radiation tumor dose was escalated for each new 5 day cycle by increments of 40-50 rads per daily session until side effects were noted or until the daily standard dose of radiation was reached.

3. Monitoring for drug response and toxicity

All patients had a pretreatment staging workup which included a detailed physical examination, liver and bone scans, a chest x-ray and when indicated, CAT scanning of the pelvis, abdomen or chest.

The irradiated palpable lesions were measured before treatment and every week during and after therapy was administered. If radiologic tests were required to measure the lesions they were repeated every 4-8 weeks.

Objective responses to treatment were defined using customary criteria: partial remission as the reduction of all measurable lesions by more than 50% of the product of the tumor's longest diameter with the perpendicular diameter on its midpoint; stabilization of disease as a reduction by less than 50% of this value and complete remission as the disappearance of a previously measurable lesion.

For each lesion responding to the combined modality of treatment a relative dose efficacy ratio (DER) was calculated; it was obtained dividing the radiation dose of the drug, by the radiation dose which brought the respective lesion to remission while doxorubicin by continuous infusion was administered. This ratio has only a relative value because it compares data obtained at different times in different studies. Nevertheless, this parameter was found useful in evaluating the impact concomitant doxorubicin infusion had on the radiation therapy efficacy.

Patients were monitored for cardiotoxicity by careful physical examinations, electrocardiograms and echocardiograms when needed. Their ventricular ejection fraction was determined initially from the patients' electrocardiogram and echocardiogram data and during the last 14 months of this study by tecnetium 99 pool MUGA scanning. Regular measurements before each new cycle were started when the cumulative dose of doxorubicin they received, reached 400 mg/m^2.

Complete blood counts and differential counts, liver function tests and BUN and creatinine values were obtained before treatment and once weekly between treatments for the first two cycles and for all cycles in which the doxorubicin dose was increased. Thereafter, they were performed only before a new cycle of therapy.

4. Pharmacokinetics of doxorubicin administered by continuous infusion.

The doxorubicin level in serum and urine was determined by a solid phase radioimmunoassay (11) using the commercially available ^{131}I doxorubicin radioimmunoassay kit (Diagnostic Bioch. San Diego, Ca.). Serum levels of doxorubicin were determined every 8 hours in the first day of doxorubicin administration and during the 48 hours following the discontinuation of doxorubicin infusion and once daily between these periods. The urinary excretion of doxorubicin was determined in samples collected by micturition every eight hours during the first 24 hours then in 24 hour samples collected daily for 10 days. The doxorubicin level was expressed in nanograms per ml.

RESULTS

1. Toxicity and optimal dose of doxorubicin by continuous infusion

In nine patients in whom escalating doses of doxorubucin were administered the limiting factor was represented by bone marrow suppression (table II) (see appendix).

At a doxorubicin dose of 15 $mg/m^2/day$ administered for five consecutive days, severe leukopenia (550±120 cells/µl) developed; its nadir was reached between the 10th and 15th day from the start of the infusion. This was accompanied by moderate to severe thrombocytopenia (35,000±40,000 platelets/µl) and moderate mucositis especially glossitis. In one occasion a patient developed sepsis which responded to antibiotic therapy. Patients' ventricular ejection fraction had only a modest decrease after exercise (by 11%). At a lower dose of doxorubicin of 13 $mg/m^2/day$ (table II) the nadir of the WBC was at 2200±750 cells/µl, that of the platelet cound at 83,000±35,000/µl, and no other side effects were noted. For this reason when administered alone, this dose was then used in all patients who entered this study. The WBC and platelet counts were back to normal by the 18th-21st day from the start of the infusion which led to a repeat of the infusion cycle every three weeks. The number of cycles administered was related to the total dose of doxorubicin the patient has received during the course of his disease. In three cases the cycles were repeated until a decrease of more than 20% of the ventricular ejection fraction after exercise was noted. This was reached at a cumulative dose of doxorubicin of 820, 860 and 850 mg/m^2 (average 840 mg/m^2). Based on this result it was decided to stop the administration of 120 hour infusion of doxorubicin at a cumulative dose of 840 mg/m^2.

2. Kinetics of doxorubicin by continuous infusion

The values of doxorubicin serum level determined by radioimmunoassay represent the amount of doxorubicin as well as that of its immediate metabolites, adriamycinol and deoxyadriamycinol aglycone, with which the anti doxorubicin antibody strongly cross reacts (11,12).

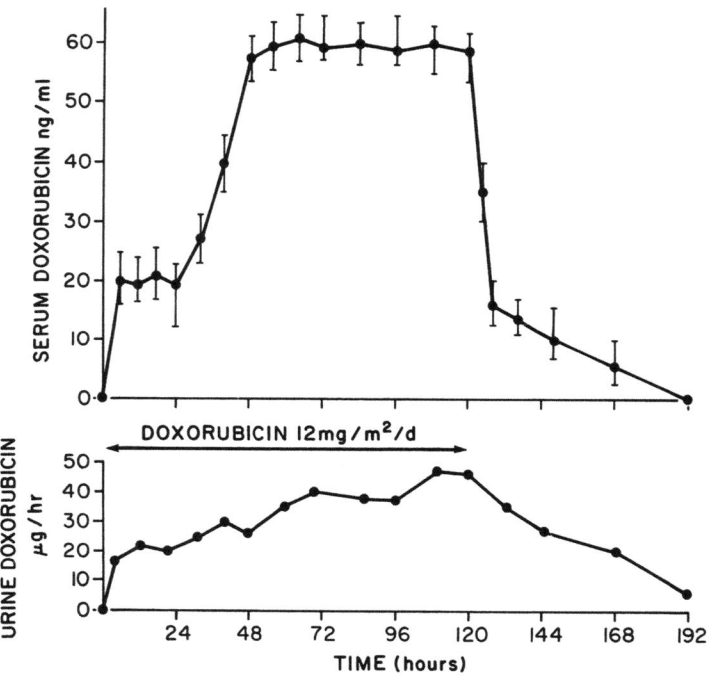

Fig. 1 Kinetics of doxorubicin administered by 5 day continuous infusion

As seen on Fig. 1 (upper graph) after starting the doxorubicin infusion at a rate of 12 mg/m^2/day, its serum level had a rapid rise to 20 ng/ml followed by a 16 hour plateau. Thereafter, there was an almost exponential increase to 60 ng/ml for another 24 hours followed by a steady plateau at that level until the infusion was terminated. The disappearance of doxorubicin and adriamycinol to 20% of its previously achieved steady level following a first order kinetics curve. This was succeeded by a much slower disappearance phase lasting 48 hours. From the disappearance curve (Fig.1) the mean half life of doxorubicin and its metabolites was calculated as 20.5 hours while the half time of the short disappearance curve was only 2.5 hours. The doxorubicin excretion in the urine (Fig. 1., bottom graph) has averaged 20 µg/hr. during the first 8 hours of infusion and has increased to 40 µg/hr. during the following 16 hours; then it continued to rise slowly up to an average of 80 µg/hr. during the last two days of infusion. Thereafter, the doxorubicin excretion has decreased first rapidly to a mean of 53 µg/hr. in the day following the termination of the infusion then at a slower pace; it was still excreted in the urine (average 9 µg/hr.) in the fourth post infusion day. The cumulative urinary excretion of doxorubicin and its immediate metabolites represented 9.7% of the total doxorubicin dose administered during the five day infusion.

3. Optimal radiation dose and toxicity of the concomitant administration of continuous doxorubicin infusion with radiation.

Based on previous data, a doxorubicin dose of 12 mg/m^2/day was chosen for administration with concurrent radiation therapy for five consecutive days every three weeks.

The effects of escalating doses of radiation (starting at 100 rads/day) were assessed on the first six patients entered in this study. No severe side effects were noted (table III, see appendix). Skin erythema and non-pitting edema developed 5 days after a daily dose of 200 rads we administered for 5 consecutive days; because of this reaction the following cases entered in this study received daily doses of 150 rads to measurable lesions in their chest and abdominal cavity and 180 rads to those in their limbs and head and neck region in cycles of five days every 3 weeks with concurrent administration of doxorubicin. At this dose local skin reaction at the site of administration of radiation consisted of hyperemia, noted after the administration of at least 1000 rads, followed by hyperpigmentation, which became apparent 4-6 weeks later; in cases receiving a total dose greater than 2500 rads, a slight degree of skin induration and thickening became apparent 3 months later.

One patient (ME) developed a more intense erythema with non-pitting edema over the irradiated area of his left lower abdominal quadrant three days after the first cycle. This patient had received 4000 rads radiation therapy to the same region 3 months prior to his concomitant radiation-doxorubicin regimen. Therapy with prednisone administered orally at a daily dose of 60 mg. abated the pain within 48 hours and resolved the erythema within 5 days. During the following cycle this patient was started on prednisone at the beginning of the cycle which prevented the recurrence of most of this local reaction.

Another patient (CT) developed pumonary fibrosis with retraction of the right hilum 4 months after receiving the combined modality treatment with the radiation delivered to the mediastinum at Rt. hilum. In this case, however, the radiation dose was much higher than that given in all other cases and it was delivered during, as well as after the adriamycin infusion, at variance with our original protocol. Alopecia was universally encountered usually after the second cycle of therapy.

Moderate mucositis manifested with dysphagia or diarrhea, due at least in part, to candida infection, developed only in 5 patients (table IV, see appendix) who had the lower third of the esophagus or the small bowel respectively included in the radiation ports. It remitted after 2 weeks therapy with mycostatin.

Patients' ventricular ejection fraction decreased only in one case at rest (by 15% of the initial value); this was the patient who received 4500 rads to the mediastinum. Only one other patient presented a reduction of the ventricular ejection fraction after exercise, following the administration of a total dose of doxorubicin of 840 mg/m^2.

Moderate bone marrow suppression developed in most of the cases (table IV) as reflected by moderate pancytopenia corresponding to that expected after the administration of doxorubicin infusion alone.

One patient (JE) with previous history of peptic ulcer had a transient GI bleeding from a duodenal ulcer area which was not included in the radiation port.

One patient (MH) with hepatoma and liver cirrhosis had a transient rise of the direct bilirubin to 2.8 mg/dl at the end of the first cycle of combined radiation doxorubicin infusion. It remitted after 5 days and did not recur with the second cycle of therapy. No other hepatic toxicity was noted in any other patient.

4. Antitumor activity of concomitant radiation therapy and doxorubicin infusion.

There were nineteen evaluable patients who received the previously described concurrent radiation-doxorubicin infusion treatment; their response to the combined modality of treatment reflected by the status of their irradiated lesions is summarized on table V and listed with more detail on Table VI (see appendix).

Three patients, all with STS, achieved (Table V) complete remission (CR). Five patients (three with STS and two with hepatoma) reached partial remission (PR); in eight, disease remained stable (SD) and in three patient's tumor progressed while receiving radiation and concomitant doxorubicin infusion.

The seven patients with metastatic adenocarcinoma of the stomach, colon and breast and the patient with metastatic transitional cell carcinoma of the bladder showed only a modest response: a reduction by less than 20% of the irradiated masses in 5 patients and just a softening of the radiated mass in three others. This stabilization of their disease lasted a mean of 8±3 weeks.

Among the four cases of hepatoma entered in this study two of them (OV and BH) with localized metastatic lesions in the Rt. lobe achieved partial remissions. In patient BH two hepatic lesions clearly defined on CT scan, decreased in size by 70% (Table VI, see appendix) after two cycles of therapy (1500 rads and 150 mg. total dose of doxorubicin by continuous infusion) while her high serum alpha fetoprotein level dropped by 85%.

Patient OV was entered in this study when she relapsed following a 6 month partial remission achieved as a result of administration of 1500 rads radiation to the liver and intermittent doxorubicin bolus injections (Fig. 2). After only 1080 rads to the liver mass and 100 mg. of doxorubicin by continuous infusion it was noted that patient's vertical liver span, on the right midclavicular line has significantly decreased from 17 to 12 cm. The size of the indurated mass of the left lobe of patient's liver, measured on liver scan, had decreased by 80% and its tenderness to palpation abated.

These partial remissions lasted 8 and a half months for OV while in BH it was interrupted by the patient's sudden death caused by the rupture of a "berry" cerebral aneurysm. At autopsy only one small mass (less than 1 cm. in diameter) was found. The other two patients with hepatocellular carcinoma accompanying cirrhosis of the liver presented only a temporary stabilization of their lesions (for 8±3 weeks).

The tumor responses of the sarcoma patients were much more remarkable. In three patients the treated lesions achieved complete responses after only 1500 rads for the Rt. arm lesion of patient EW with liposarcoma, after 2500 rads for the Rt. axillary lesion of patient BF with fibrosarcoma

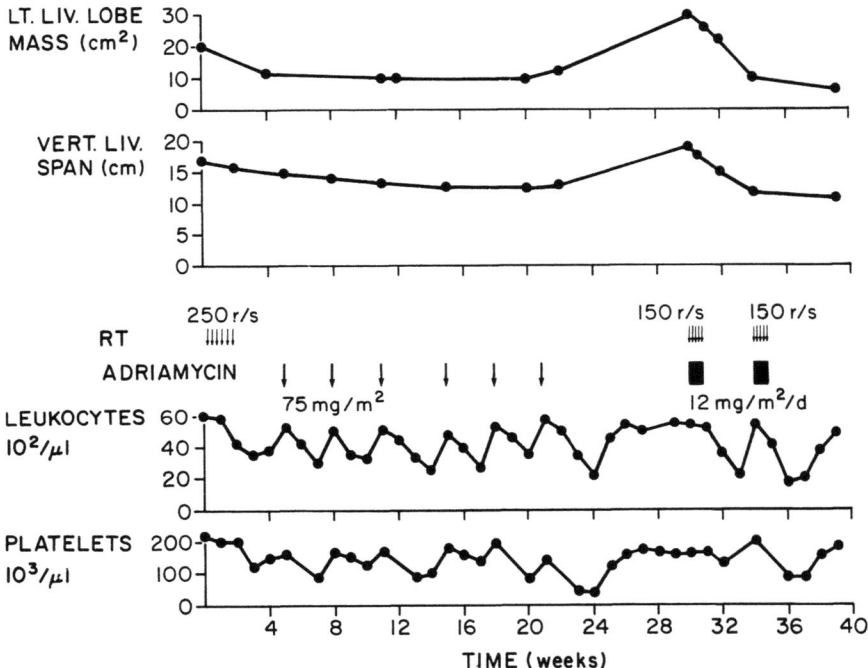

Fig. 2. Variations of different tumor and blood count parameters during, before and after the administration of continuous Adriamycin infusion with concomitant radiation therapy in a patient with hepatoma (OV).

and after 2400 rads for the Lt. iliac lesion of patient ME with undifferentiated sarcoma. In the latter CR was achieved after no response was seen following 4000 rads radiation administered alone and two chemotherapeutic regimens. Because of this prior therapy induration of the skin with subcutaneous fibrosis developed. Five needle biopsies of the indurated, fibrotic areas showed the absence of any malignant cells. In three other patients the radiated lesions achieved partial responses (decrease in size by more than 50%) after the administration of a dose of radiation significantly lower than the one causing similar responses in historical controls receiving radiation without concomitant doxorubicin infusion. The dose efficacy ratio (DER) in all responders had a mean of 2.4 ± 0.8 with a range varying between 2.1 and 3.5. In one case of leiomyosarcoma with liver metastases no significant decrease in the size of the hepatic lesions was noted after the administration of 2500 rads. Nonetheless, these lesions have not increased in size and patient has remained asymptomatic for the last 27 weeks.

Overall, the duration of the response of the radiated sarcoma lesions, which remitted after receiving combined modality of therapy, was longer than 66 ± 30 weeks for complete responders and longer than 27 ± 8 wks. for partial responders. In fact, in none of the responders, to date, a relapse at the radiated site occurred despite the low dose of radiation administered. Two of them with synovial sarcoma and liposarcoma (CT and EW) expired 8 and 10 months after therapy due to progressing metastatic lesions elsewhere in their lungs.

The 4 patients who achieved complete or partial responses of their solitary recurrent or metastatic lesions are maintaining their response 98 and 69 weeks for CR patients and 26 and 17 wks. for PR patients respectively (table VI, see appendix).

DISCUSSION

The most significant results of the present study revealed the lack of toxicity of doxorubicin administered by protracted infusion with concomitant radiation and its radiation enhancing effect in 86% of cases of sarcoma and in half of the hepatocellular carcinoma patients entered in this study. Conversely, doxorubicin infusion showed no apparent radiation enhancing effect in all of the cases of metastatic or recurrent adenocarcinoma of the colon, stomach or breast who received this regimen.

The radiation enhancing effect of doxorubicin by continuous infusion was better defined by using a new parameter, the dose efficacy ratio, calculated by dividing the dose that conventionally is delivered to achieve control of the respective lesion by the radiation dose that induced the lesion's maximum response while doxorubicin was intravenously infused. This DER varied between 2.1 and 4.0 for soft tissue sarcoma lesions indicating, in fact, that these tumors are sensitive to the combined modality treatment. This effect was even more remarkable if we keep in consideration the fact that almost all sarcoma patients have previously received doxorubicin and four of them received radiation also to the same lesions more than 3 months before they entered in this study. These data suggest that doxorubicin may indeed have a radiosensitizing effect. Such effect was previously recognized in experiments carried out in in vitro cultured chinese hamster ovary cells (14) as well as on murine, hamster (15) and human tumor cells (16). In other experiments (17) using the intestinal crypt cell assay it was noted that cell survival was reduced by the addition of doxorubicin to 70% of the survival of the crypt cells receiving radiation alone (17). There was not evidence of a time interval yielding enhanced response. Preliminary clinical studies (4,5) also indicated that doxorubicin can be safely administered in intermittent single pulses with concomitant conventional radiotherapy as well as with accelerated split dose radiation therapy or brachy radiotherapy. Side effects represented by progressive bowel fibrosis, esophagitis and dermatitis were noted in less than 30% of the cases and have been reversible except for the bowel fibrosis. Similar results and toxicity were found (18) following the administration of sequential radiotherapy and doxorubicin to patients with bronchogenic carcinoma. Favorable results with minimal morbidity were reported in the treatment of localized high grade soft tissue sarcoma (19) in that of patients with limited mesothelioma in whom radiotherapy was sequentially administered with doxorubicin alone or in combination with other drugs (20) as well as in the treatment of transitional cell carcinoma of the bladder with low dose doxorubicin (21). However, these data have generally been at variance with most of previous reports indicating increased toxicity resulting from the concomitant administration of radiation and pulses of doxorubicin i.v. A skin recall phenomenon (22) as well as severe enteritis, esophagitis (13,23) and cardiomyopathy (24) were reported. There appears to be an enhanced toxicity when doxorubicin was administered between 2 to 48 hours before radiation therapy (13). Some experimental in vivo data (25) in hepatoma H-4-II-E bearing rats have also shown that doxorubicin administered immediately after tumor irradiation failed to cause additional tumor cell kill. However, the data herein presented indicated that toxicity of the concomitant doxorubicin and radiation therapy is negligible when doxorubicin level in serum does not rise above 60 ng/ml; at this level doxorubicin appears to have an anti-neoplastic effect synergistic with radiation therapy at least on sarcomas and hepatomas.

The mechanism of the radiosensitizing effect of doxorubicin is still speculative. It has been reported (15,16,26) that doxorubicin is an inhibitor of mitochondrial and tumor cell respiration. This effect could lead to reduced oxygen consumption by cells in the outer layers of the tumor and consequently an improved oxygenation of the centrally located hypoxic cells (27). It is conceivable that radiosensitization requires a minimum steady level of doxorubicin in the serum which is readily achieved by continuous i.v. infusion of doxorubicin rather than with i.v. pulses. After the administration of the commonly recommended bolus dose of 60 mg/m^2 of doxorubicin a plasma level of 500 ng/ml is reached within 5-10 minutes (12). Thereafter, there is a biphasic disappearance of doxorubicin from plasma; its level declines to 10% in about 48 hrs. and to 5% within 120 hrs. (25). Our kinetic data shows that when an equal dose of doxorubicin is infused continuously over a period of 5 days - a steady plasma level of approximately 60 ng/ml is maintained for 100 hours (Fig.1). Starting with its 36th hour of administration this level becomes higher than the plasma level attained after i.v. pulse administration of an equal amount of doxorubicin. Its disappearance from plasma after the administration of the drug as a 120 hour infusion also had a biphasic pattern. However, the duration of the rapid disappearance phase is longer (Fig.1) than that seen after an i.v. bolus (28); this could be explained by an overlap of its disappearance from the serum into the tissues during the last hours of infusion with its release from tissues in the serum and urine. The amount of doxorubicin and of its immediate metabolites, adriamycinol and deoxyadriamycinol found in the urine represented 9.7% of the total amount infused over the 120 hours, which constitutes an increase over the usual 5% urinary excretion following i.v. bolus administration (28). This could be due to an increase of the amount of doxorubicin degraded to metabolites cleared by the kidney due to its prolonged retention by various tissues as a result of its slow and protracted administration. The increased doxorubicin retention by tissues could also explain the 24 hr. period delay in reaching maximum doxorubicin level in plasma (Fig.1).

Doxorubicin side effects have been generally minor when administered by continuous infusion in the schedule and dosage found to have the best therapeutic index (13 mg/m^2/d in five day cycles every three weeks).

Cardiac toxicity as determined by the measurement of the ventricular ejection fraction was minimal and was noted only in patients who reached a total dose of 840 mg/m² of doxorubicin of which less than half was administered as i.v. pulses prior to their entry in this study. It consisted in the reduction by 25% of less of the ventricular ejection fraction after exercise. A significant decrease of cardiac toxicity induced by doxorubicin when administered by 96 hour continuous infusion up to a dose of 60 mg/m² per cycle was also recently documented by Legha et al (7) in an extensive study, which included an analysis of endomyocardial biopsies and of cardiac radionuclide cineangiography data, as well as in two recent studies (29,30) in which doxorubicin was administered through ambulatory pump delivery systems for periods of time longer than 30 days at a daily dose lower than 4 mg/m².

The side effects of the concurrent administration of doxorubicin by continuous infusion and radiation therapy have been essentially identical with those of doxorubicin administered alone by protracted infusion. Most remarkable was the lack of detectable hepatic and cardiac toxicity.

Based on these data it can be concluded that doxorubicin at a serum level close to 60 ng/ml appears to have an enhancing effect on radiation in the treatment of most soft tissue sarcomas and of a few cases of hepatocellular carcinoma entered in this study.

This new combined modality of treatment warrants more extensive phase II and III clinical trials in the therapy of recurrent as well as primary neoplasms which respond poorly to their current treatment.

REFERENCES

1. D'Angio, G.J., Farber, S. and Maddock, C.L. Potentiation of x-ray effects by actinomycin D. Radiology 73:175-177, 1959.
2. Phillips, T.L. and Fu, K.K. Quantification of combined radiation therapy and chemotherapy effects on critical normal tissues. Cancer 37:1186-1200, 1976.
3. Byfield, J.E.: The role of radiation repair mechanism in radiation treatment failures. Cancer Chemoth. Repts. 58:527-538, 1974.
4. Byfield, J.E., Watring, W.G., Lemkin, S.R., Juillard, G.L., Hauskins, L.A., Smith, M.L. and Lagasse, L.D. Adriamycin: a useful adjuvant drug for combination radiation therapy. Proceed. Am. Assoc. Cancer Res.-Am. Soc. Cl. Onc. 16:253, 1975 (Abstr.).
5. Chan, Y.M., Byfield, J.E., Lemkin, S.R., Aronstam, E. Coincident adriamycin (A) and x-ray therapy in bronchogenic carcinoma: response and cardiotoxicity. Proceed. Am. Assoc. Cancer Res.-Am. Cl. Onc. 17-276, 1976 (Abstr.).
6. Shimoyama, M. The cytocidal action of alkylating agents and anticancer antibiotics against in vitro cultured Yoshida ascites sarcoma cells. Japanese Soc. Cancer Ther. 10:63-72, 1975.
7. Legha, S.S., Benjamin, R.S., Mackay, B., Ewer, M., Wallace, S., Valdiviesco, M., Rasmusser, SL.L., Blumenshein, G.R. and Freireich, E.J. Reduction of Doxorubicin cardiotoxicity by prolonged continuous intravenous infusion. Ann. Int. Med. 96:133-139, 1982.
8. Russell, W.O., Cohen, J., Edmonson, J.H., Enzinger, F., Hajdue, S.I., Heise, H., Martin, R.G. et al. Staging system for soft tissue sarcomas. Sem. Oncol. 8:156-159, 1981.
9. Haskell, C.M., Ellber, F.R., and Morton, D.L. Adriamycin (NSC-123127) by arterial infusion. Cancer Chemoth. Rep. 6:187-189, 1975.
10. Bjeletich, J. and Hickman, R.O. The Hickman In Dwelling Catheter. American Journal of Nursing, 80:62-65, 1980.
11. Van Vunakis, H., Langone, J.J., Riceberg, L.J. and Levine, L. Radioimmunoassays for adriamycin and daunomycin. Cancer Research 34: 2546-2550, 1974.
12. Bachur, M.R., Riggs, C.E., Green, M.R., Langone, J.J., Van Vunakis, H.

and Levine. Plasma adriamycin and daunorubicin levels by fluorescence and radioimmunoassay. Clin. Pharm. and Ther. 21:70-72, 1976.
13. Phillips, T.L., Wharam, M.D. and Margolis, L.W. Modification of radiation injury to normal tissues by chemotherapy agents. Cancer 35: 1678-1684, 1975.
14. Kimler, B.F., and Loeper, D.B. The effect of adriamycin and radiation on G cell survival. Int. J. Rad. Onc. Biol. Phys. 3:1297-1300, 1979.
15. Byfield, J.E., Lee, Y.C., Tu, L.: Molecular interactions between Adriamycin and x-ray damage in mammalian tumor cells. Int. J. Cancer 19:186-193, 1977.
16. Byfield, J.E., Lynch, M., Kulhaman, F., Chan, PYM: Cellular effects of combined Adriamycin and X irradiation in human tumor cells. Int. J. Cancer 19:194-204, 1977.
17. Ross, G.Y., Phillips, T.L. and Goldstein, L.S. The interaction of irradiation and adriamycin in intestinal crypt cells. Int. J. Rad. Oncol. Biol., Physics 5:1313-1315, 1979.
18. Ruckdeschel, J.C., Baster, D.H., McKneally, M.F., Killam, D.A. et al. Sequential radiotherapy and adriamycin in the management of bronchogenic carcinomas. The question of additive toxicity. Int. J. Rad. Onc. Biol. Phys. 3:1323-1328, 1979.
19. Blum, R.H., Greenberger, J.S., Wilson, R.E. and Carson, J.M. Feasibility of combined modality therapy for localized high-grade soft tissue sarcomas in adults. Int. J. Rad. Onc. Biol. Phys. 5:1281-1285, 1979.
20. Sinoff, C., Falkson, G., Sandison, A.G. and DeMuelenaere, G. Combined doxorubicin and radiation therapy in malignant pleural mesothelioma. Cancer Treat. Rep. 66:1605-1608, 1982.
21. Kagawa, S., Maebayashi, K., Kubokawa, K., Uyoma, T. and Moriwaki, S. Efficacy of combination therapy with intravescical instillation of doxorubicin and low dose radiation for bladder cancer. Urology 18: 479-481, 1981.
22. Cassidy, J.R. Radiation-Adriamycin interactions: preliminary clinical observations. Cancer 36:946, 1975.
23. Greco, A.A., Brereton, H.D., Kent, H., Zimbler, H., Merrill, J. and Johnson, R.E.: Adriamycin and enhanced radiation reaction in normal esophagus and skin. Ann. Int. Med. 85:294-298, 1976.
24. Rosen, G., Tefft, M., Martinez, A., Cham, W. and Murphy, M.L.: Combinnation chemotherapy and radiation therapy in the treatment of metastatic osteogenic sarcoma. Cancer 35:622-620, 1975.
25. Rowley, R., Bacharach, M., Hopkins, H.A., MacLeod, M., Ritenour, R., Moore, J.V., Looney, W.B. Adriamycin and radiation effects upon an experimental solid tumor resistant to therapy. Int. J. Rad Onc. Biol. Phys. 5:1291-1295, 1979.
26. Gonsalvez, M., Blanco, M., Hunter, J., Miko, M., Chance, D. Effects of anti-cancer agents on the respiration of isolated mitochondria and tumor cells. Europ. J. Cancer 10:567-574 1974.
27. Durand, R.E. Adriamycin: a possible indirect radiosensitizer of hypoxic tumor cells. Radiology 119:217-222, 1976.
28. Benjamin, R.S., Riggs, Jr., C.E. and Bachur, N.R. Pharmacokinetics and metabolism of adriamycin in man. Clin. Pharm. Therap. 14:592-600, 1973.
29. Lokich, J., Bothe, A., Zipoli, T., Green, R., Sonneborn, H., Paul, S. and Phillips, D. Constant infusion schedule for adriamycin: a phase I-II clinical trial of a 30 day schedule by ambulatory pump delivery system. J. Clin. Onc. 1:24-29, 1983.
30. Garnick, M.B., Weiss, G.R., Steele, Jr., G.D., Israel, M., Schade, D., Sack, M.J. and Frei, III, E. Clinical evaluation of long-term, continuous infusion doxorubicin. J. Clin. Onc. 57:1-10, 1983.

APPENDIX

Table 1. Clinical and histologic characteristics of patients receiving combined modality treatment (Doxorubicin Infusion - RT)

Pts.	Age/Sex		Diagnosis	Mets/Rec.	Stage or Grade	Perf. Stat.	Previous Rx. RT	Chemoth.
EW	69	M	Liposarcoma Rt. arm	Lung	G:3	2	-	CYVADIC Adr.:370/m²
CT	44	F	Synovial Sarc. Retroperit	Lung, Bones	G:3	3	-	CYVADIC
ME	62	M	Poorly Dif.Sarc. Retroperit	Rec.+ Pelvis	G:3	2	4000	CYVADIC VAB VI Adr.:380/m²
CJ	33	M	Leiomyosarcoma Lt. psoas	Liver	G:2	1	-	CYVADIC Adr.:260/m²
DF	68	M	Leiomyosarcoma Retroperit.	Rec.	G:3	1	-	CYVADIC Adr.:400/m²
PM	69	F	Leiomyosarcoma Duod.	Liver	G:2	0	-	CAF Adr.:160/m²
BF	72	M	Fibrosarcoma Rt. arm	Rt. axilla	G:3	1	-	CYVADIC Adr.:160/m²
SM	72	F	Adeno. Stomach	Abd. wall	S:IV	2	-	FAM x 2c Adr.:160/m²
LI	62	M	Adeno. Stomach	Abd. wall	S:IV	3	-	FAM x 6c Adr.:450/m²
JE	51	F	Adeno. Colon	Omentum	S:D	3	-	Mito + 5FU
SD	73	M	Adeno. Colon	Pelvic	S:D	2	-	5FU
SJ	30	F	Breast Rec.	Breast, Lung	S:IV	3	4500	CAF Adr.:180/m²
WJ	43	F	Breast Inflam. Recurrent		S:IV	1	1500	CMF
WH	35	F	Breast Ca. Inflam.	Bones	S:IV		-	CMF
FL	70	M	Bladder Ca. Trans.	Pelvis Bones	S:D	3	-	Cis CA Adr.:220/m²
OV	44	F	Hepatoma	Rec.	G:II	1	1500	Adr.:460/m²
BH	52	F	Hepatoma	Rec.	G:I	1	-	Adr.:300/m²
LM	68	M	Hepatoma	Omentum	G:III	1	-	---
FE	55	M	Hepatoma	Rec.+Omen.	G:II	2	-	---

S=stage; G=grade; F=5 Fluorouracil; A=Adriamycin (Doxorubicin); M=Mitomycin C; C=Cyclophosphamide; Cis=Cisplatin; V=Vincristine; DIC=Diamino Imadazole Carboxamide; B=Bleomycin.

Table II. Toxicity of Escalating Doses of Doxorubicin by Continuous Infusion

Dose	Nadir Counts		Muco-citis	Skin erythema	Ventricular ejection fraction (gated pool scanning)	
Doxorubicin (mg/m^2/dx5)	WBC (x10^3/µl)	Platelets (x10^3/µl)			Direct (nl‡:57±8%)	After exer. (nl:71±8%)
9	3.6±0.85	135±42	0	0	55%	74%
11	3.1±0.52	120±25	0	0	ND	ND
13	2.2±8.75	82±35	+	0	53%	72%
15	0.55±0.72	35±40	++	0	52%	61%

+RT = radiation therapy, ‡ = normal, **ND = not determined

Table III. Toxicity of Limited Field Radiation and Concomitant Continuous Doxorubicin Infusion

Dose			Nadir Counts		Muco-citis	Skin erythema	Ventricular ejection fraction (gated pool scanning)	
Doxorubicin (mg/m²/dx5)	+	RT[+] (rads/dx5)	WBC (x10³/µl)	PLATELETS (x10³/µl)			Direct (nl‡:57±8%)	After exer. (nl:71±8%)
12	+	100	2.1±0.8	84±27	→	→		
12	+	150	2.2±0.65	85±31	+	+	ND**	
12	+	200	2.05±0.72	80±28	++	+	ND	
						+++	53%	72%

[+]RT = radiation therapy, ‡nl = normal, **ND = not determined

Table IV. Side Effects of Doxorubicin Continuous Infusion with Concurrent Radiation

INCIDENCE

TYPE	No. of cycles (of 72) (percentage)	No. of pts. (of 20) (percentage)
Skin erythema	82.8	100
Leukopenia (WBC <1500 > 3000)	85.5	95
Thrombocytopenia (Plts.> 5x10⁴ <1x10⁵)	62.1	70
Mucositis	28.9	25
Infectious episodes (Rx at home)	24.8	50
Infectious episodes (Rx in Hosp.)	4.1	10
Cardiac arrhythmias	4.1	15
Decreased Ventricular ejection fraction after exercise	2.7	10
Cholestasis	1.3	5
G.I. bleeding	1.3	5
Pulmonary fibrosis	---	5
Alopecia	---	100

Table V. Response Rate (DER)* and Duration of Response in Patients Receiving Doxorubicin Infusion and Concomitant Radiation

Histology	No. Pts.	CR** No. Pts.	CR** Wks.	PR† No. Pts.	PR† Wks.	SD‡ No. Pts.	SD‡ Wks.	Prog. No. Pts.
Adeno Ca. of G.I. tract	4	0	–	0	–	3	8±3	1
Breast Carcinoma	3	0	–	0	–	2	8±4	1
Transitional Ca. of Bladder	1	0	–	0	–	0	–	1
Hepatocellular Ca.	4	0	–	2	28±6	2	8±3	0
Soft Tissue Sarcomas	7	3 (2.5)*	66±30	3 (2.3)⁺	27±9	1	27	0
Total		3	66±30	5	27±8	8	12±9	3

*DER = Dose Efficacy Ratio listed in parenthesis: standard RT dose when administered alone/RT dose inducing same response in the presence of doxorubicin infusion; **CR = complete response; †PR = partial response; ‡SD = stable disease.

Table VI. Effects of Concomitant Radiotherapy-Continuous Infusion Doxorubicin on Treated Malignant Lesions

Pt's Initials	Diagnosis	Administered Rx Doxorubicin (total mg/m^2)	RT* (total rads)	Measurable Lesions (cm^2) Before Rx	After Rx	DER	Response Type	Duration (wks)
EW	Liposarcoma	120	Rt.arm:1600 **RUL:2000	48	0 24	3.7	CR	33
CT	Synovial Sarcoma	240	Rt.Hilum:4500 +LUQ:2370	12 80	2 6	4 2.5	PR	38
ME	Poorly Diff. Sarc.	120	‡LLQ:1500	84	15(-bx)	3.6	CR	98+
CJ	Leiomyosarcoma	180	Liver:2250	18	14	--	SD	27+
DF	Leiomyosarcoma	240	Retrop.:2000	40	18	3	PR	26+
PM	Leiomyosarcoma	180	Liver:1500	20	8	3.5	PR	17+
BF	Fibrosarcoma	240	Rt.axilla:2800	56	0	2.14	CR	69+
SM	Adeno Ca. Stomach	200	Abd.wall:1920	315	280	--	SD	11
LI	Adeno Ca. Stomach	180	Abd.wall:2340	20	18	--	SD	9
JE	Adeno Ca. Colon	120	Pelvis: 750 Liver: 1320	36 8	36 6	--	Prog.	--
SD	Adeno Ca. Colon	120	Pelvis:2300	20	20	--	SD	5
SJ	Breast Ca.(rec. + mets)	120	Chest wall:	120	120	--	Prog.	--
WJ	Breast Ca.(inflam.)	180	Breast:	42	30	--	SD	12
WH	Breast Ca.(recur.)	60	Breast:	35	35	--	SD	5
FL	Transit.Cell Ca.Bladder	180	Lt. thigh:4860	150	140	--	Prog.	--
OV	Hepatoma	60	Lt.lobe Liver:1080	30	12	2.3	PR	35
BH	Hepatoma	120	Rt.lobe Liver:1500	12	4	1.6	PR	22+*
LM	Hepatoma	180	Whole Liver:2150	36	30	--	SD	5
FE	Hepatoma	140	Whole Liver:2520	16(V.sp)	15	--	SD	12

Legend – same as per table V plus: *RT = radiation therapy; **RUL = right upper lobe of the lung; +LUQ = Left upper quadrant of the abdomen; ‡LLQ = left lower quadrant of the abdomen.

CIS PLATIN BY CONTINUOUS INFUSION WITH CONCURRENT RADIATION IN MALIGNANT TUMORS (A PHASE I-II STUDY)

C. J. Rosenthal, M. Rotman, K. Choi and J. Sand

Department of Medicine, Division of Medical Oncology
Department of Radiation Oncology
Downstate Medical Center-SUNY, Brooklyn, NY

Cis-Diamminedichloroplatinum II (Cis-DDP or Cisplatin) the first of the group of platinum coordination complexes used effectively against some human malignancies, was found, since 1974, to potentiate irradiation effects on experimental tumors (1). Since 1977 Douple and Richmond (2-4) have concluded, based on a number of experimental studies on E coli bacterium (5), V-79 chinese hamster cells, mouse mammary adenocarcinoma, etc. that cisplatin sensitizes hypoxic cells to radiation, increases DNA susceptability to radiation damage and at the same time inhibits the radiation induced repair of potentially lethally damaged DNA. A convincing experimental demonstration of the radiosensitizing effect of cisplatin was obtained by Kyriazis et al (6) on a human bladder carcinoma implanted in nude mice. The mice receiving both modalities of treatment had a statistical significantly longer response and survival than those receiving radiation alone. Other experimental data suggested that the Cisplatin-Radiation interaction was enhanced when radiation was administered in multiple fractions (7) and that cisplatin could be possibly more effective when administered by continuous infusion because even so it is a phase cycle nonspecific drug, DDP acts preferentially on the G_1 phase (8). On clinical grounds preliminary studies (9-11) of cisplatin-radiation combination have used only bolus i.v. administration of the drug. While indicating a favorable trend they have been inconclusive in proving its radiosensitizing effect.

For this reason we first performed a phase I study whose objective was to find out the dose and the duration of the cisplatin infusion with best therapeutic ratio. This was followed by a preliminary phase II study whose objective was to establish the effect of cisplatin continuous infusion administered concurrently with radiation therapy on several squamous cell carcinoma malignant tumors chosen as target because of their high local recurrence rate after radiation therapy alone (12).

METHODS

Cis platin was administered by continuous infusion in an ambulatory setup. The total daily dose was delivered slowly at a constant rate by a battery activated pump (Cormed-Medina, N.Y.) attached to an indwelling Hickman's catheter inserted in the patient's subclavian vein and positioned with its tip in the superior vena cava.

The regimen of cisplatin continuous infusion with concomitant irradiation found in the phase I study to have the best therapeutic ratio, was administered concurrently with radiation therapy to 20 patients with unresectable or metastatic squamous cell carcinoma: 8 of the lung, 7 of the head and neck and 5 of the esophagus. Their age range between 25 and 75 y.o. There were 6 females and 14 males. They all had prior antineoplastic therapy but none received prior radiation to the lesion targeted for therapy in this study.

Radiation therapy was administered either as a single dose of 180-200 cGy/session to 8 patients or as a 110 cGy/session twice daily (hyperfractionation) to 12 patients. The two methods were equally distributed among the three groups of patients.

RESULTS

Table 1: Limiting Factors in the Administration of Cis-Platin by Continuous I.V. Infusion

Dose	Nausea + Vomiting	Leukopenia (10^3 cells/μl at nadir/d)	Thrombocytopenia (10^3 cells/μl at nadir/d)	↑ Ser. Creat. mg/dl/d
1 mg/m^2/d	0	4.9±4/28±5	145±12/28±5	0
3 mg/m^2/d	0	3.1±.3/27±4	108±16/29±5	0
5 mg/m^2/d	0	2.3±.6/20±5	85±12/19±6	0
7 mg/m^2/d	+	2.1±.4/18±3	79±21/18±5	0.2/19
9 mg/m^2/d	++	1.7±.3/17±6	75±18/17±7	0.2/17

As shown on table 1 the dose limiting factor of the continuous cisplatin infusion was, in most of the patients, represented by gastrointestinal discomfort when the daily dose was increased above 5 mg/m^2. Instead bone marrow suppression was the factor limiting the duration of cisplatin administration; it occurred after 19-20 days at a dose of 5 mg/m^2/d. For this reason, for the phase II study, a dose of 5 mg/m^2/day cisplatin was administered for 14 days every 28 days until complete remission (CR) or the conventional total dose of radiation were reached.

Table 2: Results of the Concomitant Combined Modality Regimen

Tumor Site	Total #pts.	CR #pts.	CR duration wks.	PR #pts.	PR duration wks.	SD #pts.
Lung	8	3	46±13	3	17±8	2
Head and Neck	7	5	55±11	2	19±7	0
Esophagus	5	3	60±21	2	23±9	0
TOTAL	20	11		7		2

As seen on table 2 there have been eleven CR and 7 PR for a total response rate of 90%. All head and neck tumors responded to this treatment. Unfortunately, the duration of this response was short ranging between 29-41 weeks. Only 4 patients are still alive. 60% of these patients succumbed because of progression of the disease at metastatic sites while 40% had local recurrences.

Table 3: Side Effects of the Radiation-Concomitant Cisplatin Infusion Regimen

	Patients (out of 20)		Cycles (out of 50)	
	Number	Percentage	Number	Percentage
Nausea/Vomiting	4	20	6	12
Anorexia	6	30	24	48
Dysphagia/Mucositis	10	50	20	40
Leukopenia (<3,000 >1,500)	5	25	7	14
Thrombocytopenia (<100,000 >50,000)	3	15	4	8
Radiation pneumonitis	2	25		
Renal Toxicity	0	0	0	0

Only minor side effects were encountered (table 3). Dysphagia due to mucositis was present in 50% of cases including all those that received radiation to the mediastinum. It was never severe to require enteral tube feeding. Moderate bone marrow suppression occurred after less than 15% of all cycles. Gastrointestinal discomfort was rarely encountered. Two patients with lung Ca. developed radiation pneumonitis 3 and 5 weeks after the treatment; it rapidly resolved with steroid therapy.

DISCUSSION

The most important result of this study indicate that cisplatin by continuous infusion was well tolerated up to a dose of 5 mg/m^2/day for maximum 18 consecutive days with the concomitant administration of radiation therapy and that this concurrent combined modality regimen led to a very high response rate especially in patients with recurrent unresectable head and neck squamous cell carcinoma. This response rate of 90% is clearly superior to that seen in historical controls receiving radiation alone. Especially remarkable was the 75% complete remission rate seen in patients with head and neck carcinoma. The dose of radiation at which remission was noted in most of the cases of lung and head and neck carcinoma was 30-40% lower than the total radiation dose conventionally administered to these tumors in the absence of cisplatin which suggests that cisplatin has a radiosensitizing effect on these human malignancies.

All previous clinical studies of concomitant cisplatin-radiation therapy have used cisplatin administered by i.v. bolus generally in relatively high doses. A 100% response rate was noted in patients with bladder carcinoma (10) while the response rate was somewhat lower in cases of squamous cell carcinoma (9,11). While our study did not compare the radiosensitizing effects of the continuous infusion cisplatin with these resulting from its i.v. bolus administration one can speculate that the infusion could be more effective in view of the fact that the radiosensitizing state is a cellular condition that develops gradually over 24 hours or more after a radiation exposure (13).

It should also be emphasized that a more conclusive clinical evidence for a synergistic effect of cisplatin and concomitant irradiation could come only from a prospective study with several arms comparing radiation therapy alone with radiation-cisplatin continuous infusion and radiation cisplatin by i.v. bolus.

The results herein reported (like those of previous studies) have also indicated that despite the high remission rate brought by the concurrent administration of Cisplatin with radiation the duration of this response was disappointedly short: 36±12 weeks; in 40% of cases local and regional recurrence occurred at the radiated site. This indicates that the concurrent administration of RT-Cisplatin while likely more effective than radiation therapy alone does not achieve in the current schedule of administration the erradication of malignant cells. Increasing the cisplatinum dose without causing severe G.I. side effects with the help of new antiemetics and using more than two fractions of radiation daily are new areas of investigation we currently explore with the hope that the radiosensitizing effect of cisplatin can be modulated to the point of achieving permanent control of the malignant lesions so treated.

REFERENCES

1. Wodinsky, I., Swiniarski, J., Kensler, C.I., Venditi, J.M. Canc. Treat. Rep. 4:73-76, 1974.
2. Douple, E.B., Richmond, R.C., Logan, M.E. J. Clin. Hematol. Oncol. 7: 585-603, 1977.
3. Douple, E.B., Richmond, R.C. Br. J. Cancer 37: Suppl. III 98-102, 1978.
4. Douple, E.B., Richmond, R.C. Int. J. Radiat. Onc. Biol. Phys 5:1369-1372, 1979.
5. Richmond, R.C., Zimbrick, J.D., Hykes, D.L. Radiat. Res. 71:447-560, 1977.
6. Kyriazis, A.P., Yagoda, A., Kereiaskes, J.G., Kiriazis, A.A., Whitmore W. Cancer 52:452-457, 1983.
7. Dritschilo, A., Piro, A.J., Kelman, A.D. Int. J. Radiat. Oncol. Biol. Phys. 5:1345-1349, 1979.
8. Sigdestad, C.P., Grdina, D.J., Peters, L.F. Proc. Am. Assoc. Canc. Res. 20:178, 1979.
9. Reimer, R.R., Gahbauer, R., Bukowski, R.M. Cancer Treat. Rep. 6:219-222, 1981.
10. Soloway, M.S., Ikard, M., Scheinberg, M., Evans, J. J. Urol. 128:1031-1033, 1982.
11. Pinedo, H.M., Karim, A.B., vanVliet, W.H. Canc. Res. Clin. Oncol.
12. Cox, J.D. Radiology 128:205, 1978.
13. Byfield, J.E., Calabro-Jones, P., Klesak, I. and Kulhanian F. Inter. J. Radiation Onc. Biol. Phys. 8:1923-1933, 1982.

SECTION II: SYNERGISTIC ANTINEOPLASTIC EFFECTS OF RADIATION THERAPY
AND CHEMOTHERAPY BY CONTINUOUS INFUSION

    B. Clinical Studies

COMBINATION OF RADIATION WITH CONCOMITANT CONTINUOUS ADRIAMYCIN INFUSION IN
A PATIENT WITH A PARTIALLY EXCISED PLEOMORPHIC SOFT TISSUE SARCOMA OF THE
LOWER EXTREMITY

S. Turner*, R. Shetty*, H. Gandhi+, A. Latyshevsky++,
J. Korzis**, and R. Yaes+

*Division of Radiation Oncology, +Division of Medical Oncology
++Department of Surgery, **Department of Pathology, Long Island
College Hospital, *+Department of Radiation Oncology, Downstate
Medical Center

It has been known for many years from the observation of acute and chronic normal tissue damage in patients receiving multimodality therapy that the effects of radiation may be enhanced in the presence of many chemotherapeutic agents (1,2,3). This observation has naturally led to many attempts to use these agents to sensitize tumors to radiation and thus obtain better local control and survival than would be possible with radiation alone. Ideally the agent used should itself be active against the tumor being treated in order to obtain the benefit of direct cell killing as well as that of radiation sensitization. The usual method of delivery of the chemotherapeutic agents has been by IV bolus, however, Byfield et al (4) and Rotman, Rosenthal et al (5) among others have done extensive investigation of the use of concomitant continuous IV infusion chemotherapy for radiation sensitization. The advantage of this mode of delivery as IV bolus is that the tumor is exposed to the chemotherapeutic agent for a longer period of time and the toxicity is diminished although a higher dose of agent is used (6).

The surgical treatment of soft tissue sarcoma of an extremity is total excision, if this is possible, in combination with adjuvant radiotherapy, or amputation (7,8). If all gross disease cannot be removed, the prognosis is dismal even if high dose post-operative radiation is given.

Adriamycin (Doxorubicin) is considered to be one of the most active single agents against soft tissue sarcomas and is used in most multiagent chemotherapy regimens for these tumors (7). Adriamycin has also been shown to be a potent radiation sensitizer in both in vitro and in vivo experiments and also in clinical studies. Watring and Byfield et al (9) have shown that Adriamycin inhibits the emzymatic repair of radiation-induced single strand breaks in DNA. Durand (10) has shown that Adriamycin selectively sensitizes hypoxic cells. Shaeffer and El-Mahdi (11) have shown in a mouse lung tumor model, that the sensitizing effect of Adriamycin is greatest if it is given during or just after radiation rather than before.

In 1975, Wiley et al (12) used an intra-arterial infusion of Actinomycin D as a radiation sensitizer in the treatment of unresectable retroperitoneal soft tissue sarcomas and in 1974, Watring, Byfield et al (9) used bolus Adriamycin as a radiation sensitizer in the treatment of advanced ovarian tumors. Sordillo et al (13) obtained 6 complete responses and 12 partial responses in 29 patients with soft tissue sarcomas treated with low dose Adriamycin given 90 minutes before radiation. In patients with multiple metastatic lesions, no response was noted in lesions outside the radiation field, to the low dose Adriamycin alone. When these lesions were subsequently irradiated without Adriamycin, much less of a response was noted. Rotman, Rosenthal et al (14) obtained a complete response in 3 of 7 patients with soft tissue sarcoma treated with radiation and continuous Adriamycin infusion. We wish to report on a patient that we have treated with the regimen developed by Rotman and Rosenthal.

CASE REPORT

Patient R.L. is a 60 year old longshoreman who presented in January of 1984 with a 6 month history of a painless right popliteal mass, increasing in size. His past medical history was remarkable for alveolar cell carcinoma of the lung treated in 1973 with right upper lobectomy and post-operative radiation. The only positive physical finding was a large firm non-tender

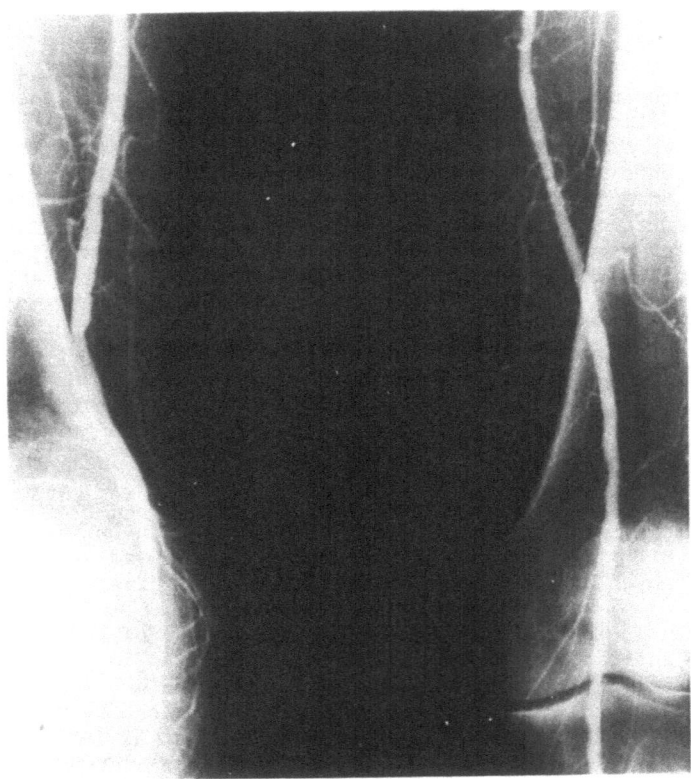

Fig. 1. Pre-operative bilateral femoral arteriogram.

mass in the right popliteal fossa. Work-up included bilateral femoral arteriogram (Fig. 1) which showed a right popliteal mass displacing the popliteal artery posteromedially. There were tumor vessels as well as a tumor blush. Sonogram and CT scan (Fig. 2) demonstrated a 25 x 8 x 8 cm. complex mass involving the muscles in the popliteal fossa. There was no evidence of metastases on chest x-ray, CT scan of the pelvis and liver scan.

The patient underwent surgical exploration of the right popliteal fossa, but because of involvement of the popliteal vessels and nerves, only partial removal of the tumor was performed. The mass removed was reddish brown to grey hemorrhagic and cystic in appearance. Histological diagnosis was grade

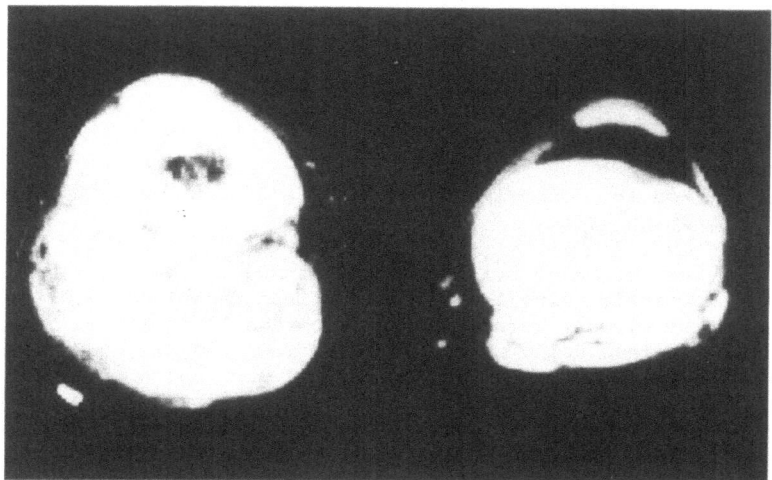

Fig. 2. Pre-operative bilateral lower extremity CT scan.

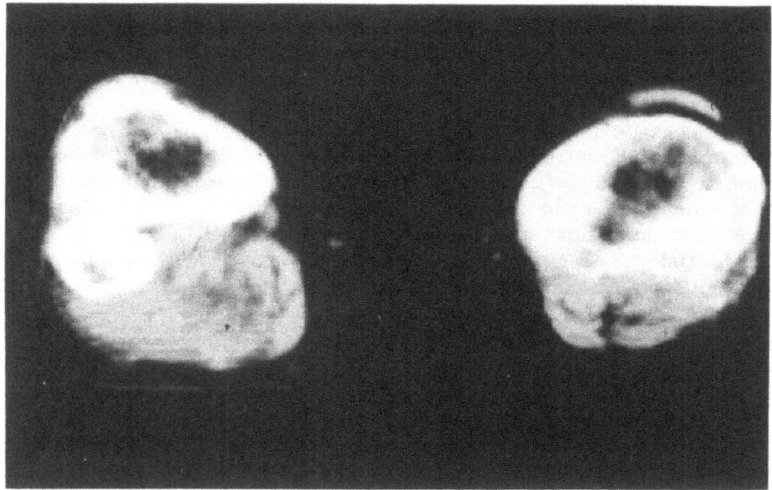

Fig. 3. Post-operative bilateral lower extremity CT scan.

III sarcoma, most likely angiosarcoma. Post-operative CT scan showed a residual right popliteal mass measuring 16 x 8 x 8 cm. (Fig. 3). The patient refused further surgery for this T-3 grade 3 Stage IV soft tissue sarcoma. It was, therefore, decided to treat him with radiation and concomitant Adriamycin infusion. (See Fig. 4 for schema).

SCHEMA

I   COMBINED TREATMENT

      Days 1-5 RADIATION 180 rads per day.
      Days 1-5 ADRIAMYCIN 12mg/M2/day continuous IV infusion.
      Days 6-21 NO TREATMENT.
      Repeat cycle x 6.

II   ADRIAMYCIN ALONE

      Days 1-5 ADRIAMYCIN 12mg/M2/day continuous IV infusion.
      Days 6-21 NO TREATMENT.
      Repeat cycle x 3.

Fig. 4. Schema for concomitant radiation and continuous IV Adriamycin infusion.

Radiation was delivered utilizing Co60 at 80 cm. SSD through parallel opposed lateral portals using the shrinking field technique (7,8) from 2/6 to 5/25/84. He received a total dose of 5400 rads TD in 30 fractions over 94 elapsed days, with a daily fraction of 180 rads per day, 5 days per week for a total dose of 900 rads per week. During each course of radiation, the patient received 12mg/M2/day of Adriamycin as a continuous infusion through a Hickman Catheter for a total dose of 60mg/M2 of Adriamycin per treatment cycle. The treatment was given in 6 weekly cycles, the cycles being separated by 2 week rest periods. By the beginning of the third cycle, the mass was no longer palpable, but a CT scan at 3240 rads revealed an ill defined posterior popliteal fossa soft tissue density. Because of this it was decided to give 3 additional cycles of Adriamycin infusion alone after completion of radiotherapy to a total overall dose of Adriamycin of 540 mg/M2. The posterior popliteal mass was still seen on a follow-up CT scan on 7/11/84 (Fig. 5), one and a half months after radiation therapy, therefore, the area was re-explored and biopsied on 8/11/84. No residual tumor was seen and pathology showed adipose tissue with hypervascularity and an inflammatory infiltrate. A repeat CT scan on 1/10/85 showed no change from 7/11/84.

The patient was last seen in follow-up on 6/5/85, seventeen months after diagnosis and had no clinical evidence of local recurrence or metastases. His range of motion at the right knee joint was excellent. His left ventricular ejection fraction has remained normal. There has been no clinical or radiological evidence of radiation recall in his previously irradiated lung during or after his treatment with Adriamycin.

DISCUSSION

In this case, the Adriamycin appears to have had a definite potentiating effect on the radiation. 5400 rads in 30 fractions over 94 days corresponds to an NSD (Nominal Standard Dose) of 1448 rets. The dose of radiation alone needed for local control of a large soft tissue sarcoma is at least 8000 rads in 8 weeks or an NSD of 2133 rets. Using a tourniquet technique (15) for extremity lesions to protect normal tissue, Suit has used doses as high as 14,000 rads in 14 fractions over 43 days. Concomitant Adriamycin infusion may substantially decrease the dose of radiation necessary for local control of soft tissue sarcoma. However, further follow-up will be necessary to see if the complete response in this patient will actually result in long term disease-free survival.

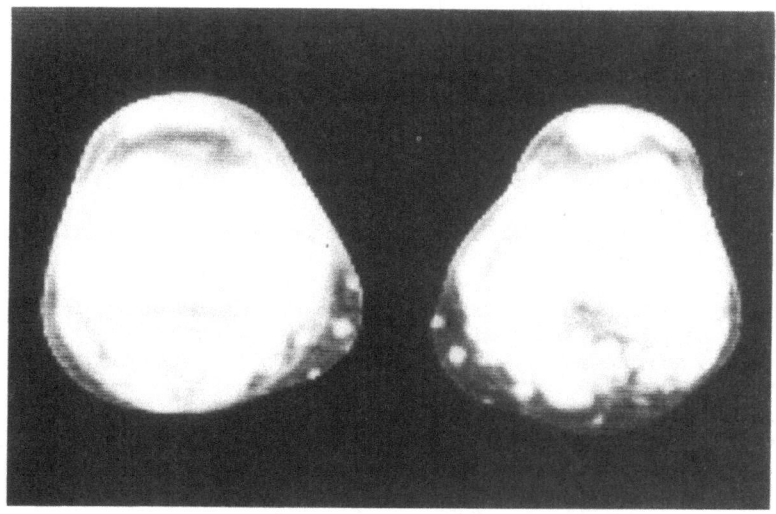

Fig. 5. Bilateral lower extremity CT scan January, 1985, 8 months after end of radiotherapy.

REFERENCES

1. Vermund, H.
   Enhancement of Radiation Effects by Chemotherapy.
   Acta Radiol. Suppl. 311 pg. 1-78, 1971.
2. Phillips, T. and Fu, K.
   Quantification of Combined Radiation Therapy and Chemotherapy Effects on Critical Normal Tissues.
   Cancer 37 pg. 1186-1200, 1976.
3. Tefft, M., Lattin, P., Jereb, B., Cham, W., Ghavimi, F., Rosen, G., Exelby, P., Marcove, R., Murphy, M., and D'Angio, G.
   Acute and Late Effects on Normal Tissues Following Combined Chemo-Radiotherapy for Childhood Rhabdomyosarcoma and Ewing's Sarcoma.
   Cancer 37 pg. 1201-1213, 1976.
4. Byfield, J., Barone, R., Mendelsohn, J., Frankel, S., Quinol, L., Sharp, T., and Seagren, S.
   Infusional 5-Fluorouracil and X-ray Therapy for Non Resectable Esophageal Cancer.
   Cancer 45 pg. 703-708, 1980.
5. Rosenthal, C., Bhutiani, I., Choi, K., and Rotman, M.
   Low Dose Adriamycin by Continuous Intravenous Infusion With and Without Concomitant Radiation Therapy.
   Proceedings of the 13th International Congress Against Cancer 1983.
   Abstract #3446 pg. 603, 1983.
6. Lokich, J.
   Infusion Chemotherapy for Cancer.
   Current Concepts in Oncology Vol. 6 #1 pg. 3-8, Spring, 1984.
7. Rosenberg, S., Suit, H. and Baker, L.
   Sarcomas of Soft Tissues
   "Cancer Principles and Practice of Oncology", Second Edition.
   Devita, V., Hellman, S., and Rosenberg, S., Editors.
   Published by J. B. Lippincott Co., Philadelphia, 1985.

8. Suit, H., Russel, W. and Martin, R.
   Sarcoma of Soft Tissue: Clinical and Histopathologic Parameters and Response to Treatment.
   Cancer 35 pg. 1478-1483, 1975.
9. Watring, W., Byfield, J., Lagasse, L., Lee, Y., Juilard, G., Jacobs, M., and Smith, M.
   Combination Adriamycin and Radiation Therapy in Gynecologic Cancers.
   Gynecologic Oncology 2 pg. 518-526, 1974.
10. Durand, R.
    Adriamycin: A Possible Indirect Radiosensitizer of Hypoxic Tumor Cells.
    Radiology 119 pg. 217-222, 1976.
11. Shaeffer, J. and El-Mahdi, A.
    Combination Adriamycin-Radiation Treatment of Pulmonary Tumors in Mice.
    Oncology 38 pg. 35-38, 1981.
12. Wiley, A., Wirtanen, G., Joo, P., Ansfield, F., Ramirez, G., Davis, H., and Vermund, H.
    Clinical and Theoretical Aspects of the Treatment of Surgically Unresectable Retroperitoneal Malignancy with Combined Intra-arterial Actinomycin D and Radiotherapy.
    Cancer 36 pg. 107-122, 1975.
13. Sordillo, P., Magill, G., Schauer, P., Vikram, B., Kim, J.H. and Hilaris, B.
    Preliminary Trial of Combination Therapy with Adriamycin and Radiation in Sarcomas and Other Malignant Tumors.
    Journal of Surgical Oncology 21 pg. 23-26, 1982.
14. M. Rotman
    Personal Communication (See also papers by M. Rotman and by C. J. Rosenthal in this volume).
15. Suit, H. and Lindberg, R.
    Radiation Therapy Administered Under Conditions of Tourniquet-Induced Local Tissue Hypoxia.
    Am. J. Roentgenol 102 pg. 27-37, 1968.

TREATMENT OF RECURRENT CARCINOMA OF THE PARANASAL SINUSES USING
CONCOMITANT INFUSION CIS-PLATINUM AND RADIATION THERAPY

M. Rotman[1], K. Choi[1], S. Isaacson[1], J. Rosenthal[2],
A. Braverman[2], and J. Marti[3]

SUNY, Downstate Medical Center, [1]Dept. of Radiation Oncology
[2]Medicine, [3]Surgery, 450 Clarkson Avenue, Brooklyn, New York

INTRODUCTION

Radiotherapeutic control of squamous cell carcinoma extending to or destroying adjacent bone is unlikely without surgical resection. Successful treatment even with combined modality therapy is unlikely when tumor extends beyond the confines of the adjacent bone. Such is the case in advanced carcinomas of the paranasal sinuses. Unfortunately, paranasal sinus tumors often remain asymptomatic until advanced with an average delay of six months between the onset of symptoms and the start of treatment. (1)

Over the past several decades various therapies have been tried to improve local control with only modest success. Combined modality primary treatment of advanced tumors of the maxillary sinus yields a local control of only 50%. (2) Results are poorer still with patients who have failed previous treatment. (3) Attempts to improve delivery of chemotherapy by intraarterial infusion alone or combined with radiotherapy and or surgery have yielded inconsistent results.

In recent years hyperfraction and accelerated fractionation have shown promise in improving control of advanced head and neck cancers, while sparing normal tissues. (12,13,14) Cis-platinum has been shown to be one of the more effective chemotherapeutic agents in terms of tumor response in epithelial malignancies. Radiation with concomitant infusion chemotherapy have proved more effective in treating several epithelial malignancies than radiation alone. It therefore seemed reasonable to combine hyperfractionated radiation with concomitant infusion cis-platinurm chemotherapy in an attempt to obtain still better control rates. This paper presents the early results of a pilot study using this combination in the treatment of advanced and recurrent carcinoma of the paranasal sinuses as well as a case that showed response of large fixed cervical nodes having metastasized from a previously treated carcinoma of the maxillary antrum.

MATERIAL AND METHODS

Two patients presented with extensive recurrent maxillo-ethmoid sinus disease and one patient with fixed cervical nodes from a previously resected paranasal sinus tumor were treated with radiation therapy and simultaneous chemotherapy. Chemotherapy was delivered by a continuous intravenous infusion concomitantly with irradiation at 2 weeks intervals (weeks 1 and

2, 5 and 6, 9 and 10). Cis-platinum was given at a dosage of 5 ml/m^2/day and was delivered continuously via a central vein catheter for two weeks. Cobalt 60 teletherapy units delivered 220 to 250 rads per day in two 110 to 125 rad fractions spaced 4 hours apart. The total dose each in each 2 week period was 2200 to 2500 rads.

CASE REPORT

A.E. is an 81 y.o black male presented with recurrent massive bilateral cervical nodes 13 weeks following treatment for a T4N0M0 destructive lesion of the left maxillary sinus that extended medially into the nasal fosae, anteriorly into the soft tissue of the face, laterally into the infratempral fosae, and inferiorly into the alveolar ridge with considerable bone distruction. Ethmoid involvement as well as the involvement of the inferior orbital plate was present. The nodes were 5cm x 4cm in diameter and fixed in the contralateral neck and 2 x 2cm in the ipsilateral neck. Therapy consisted of maxillectomy and post-operative irradiation of 6000 rads. A massive bilateral nodal recurrence was noted at 13 weeks. The total radiation dose was 6750 rads in 3 courses of treatment. The cis-platinum was initially given at 12mg/day and then reduced to 9mg/day (5mg/m^2/day) because of vomiting. The patient had achieved complete regression within one month after completion of treatment and remains NED at 19 months. Erythemal mucosal reaction was noted during the 2nd and 3rd courses of treatment.

CASE REPORT

Y.W. is a 66 y.o oriental male with a T4N0M0 carcinoma of the inferior portion of the left maxillary sinus with extension into the hard palate involving approximately one half of the hard palate. The midline was not crossed. The initial treatment was a modified maxillectomy. Residual rapidly growing tumor was noticed in the cheek and soft palate 2 weeks post-operatively. The total radiation dose was 6126 rads. Cis-platinum was given at 9mmg/day. The patient tolerated treatment well and developed mild mucositis. A complete regression of the recurrent mass was achieved shortly after the completion of treatment. Patient died with disease of recurrence in the neck in 4 months after completion with the primary controlled.

CASE REPORT

C.W. is a 71 y.o black male with extensive tumor involving the right maxillary antrum extending medially to involve the entire nasal chamber and a portion of the left maxillary sinus. Tumor extended superiorly into the right orbit causing exophthalos and posteriorly into the nasopharynx. The lesion was staged as a T4N0M0. The dose of radiation was 6960 rads. Cis-platinum dosage was 12.5mg/day. The patient developed mild mucositis and remained NED until he died of bone metastases 5 months after completion of treatment.

Patients were treated by the above described technique.

RESULTS

No complications arose from the techniques employed other than mucosal erythema or mild degree of mucositis. Response to therapy was made on the basis of post mortem examination and in one case repeat CAT scan and physical examination. Local control was achieved in all four patients with three of the four obtaining ablation of massive disease. Satisfactory palliation was accomplished in each patient but three of the four patients eventually expired of systemic disease. One (C.W.) with diffuse bone metastases and a second (A.R.) progressive cachexia and pneumonia and a third patient with failure in the neck. All three patients had disease controlled in the treated area at the time of death.

DISCUSSION

Fletcher and other have shown that in head and neck epithelial cancers, the tumor burden increases the dose required for control. (15) Efforts to increase the tumor dose to required amounts have been associated with both acute and late effects.

Withers has reported that the therapeutic ratio should improve as the size of the dose/fraction decreases. (16) Twice-a-day fractionation schedules have the potential for improving the therapeutic ratio. The mechanism appears to be related to the repair of sublethal damage and regeneration of normal cells compared to redistribution and reoxygenation of tumor cells. Clinical evidence sequela are not increased and relative control rates are better than expected with conventional radiation. (12,13,14,17)

As early as 1958 Heidelberger et al. (18) had shown that either irradiation or chemotherapy would be curative for certain rodent tumors if both treatments were combined. Other authors had also shown this effect. (19,20) The combination of chemotherapy and radiation simultaneously may also improve the therapeutic ratio.

Cis-platinum, a drug with known cytotoxic activity in squamous cell carcinoma of the head and neck, was used as a ration sensitizer. Cis-platinum increases radiation sensitivity particularly under hypoxic or anoxic conditions; conditions that are commonly found in advanced squamous cell carcinoma of the head and neck. (21) Proposed mechanisms of action include inhibition of sublethal damage repair and potential lethal damage repair, conversion of lesions into chromosomal abberations cell cycle pertubation and reaction with SH groups. (22)

Infusion therapy appears to be potentially more advantagous than bolus injection. Prolonged infusion allows exposure of all cells in the chemosensitive synthetic phase of the cell cycle. It also allows cells, which subsequently enter the cell cycle, to be exposed since only a random portion of cells are in cycle at any given time.

Continuous intravenous infusion of cis-platinum has been used in higher doses than in the present study with acceptable toxicity. (23) However, those patients were not receiving concomitant radiation. Prior treatment may exacerbate the inherent toxicity of chemotherapeutic regimen using Cis-platinum. (24) In an effort to minimize the additive toxicity of concomitant therapy the dose of Cis-platinum was kept relatively small.

CONCLUSION

In conclusion, a split course of twice-a-day radiation with concomitant cis-platinum infusion yielded a satisfactory local control rate with minimal complications, suggesting an improvement of the therapeutic ratio. The optimal dosage of infusion chemotherapy and concomitant radiation therapy is yet to be determined. Chemotherapy dosage as well as time dose relationships used in delivering radiation therapy are being evaluated. At present, there is no effective conventional treatment modality for recurrent paranasal sinus carcinoma. Retreatment of previously irradiated areas are rarely attempted because of uniformly poor results. The combination low dose infusion chemotherapy and twice daily radiation may promise a solution to this problem.

SUMMARY

Three cases of locally advanced or recurrent carcinoma of the paranasal sinuses were treated with a course of hyperfractionated radiation and concomitant Cis-platinum infusion. 5 mg/m^2/day Cis-platinum was delivered by continuous 24 hour infusion of 2 weeks duration via a central vein catheter. Concurrently patients received radiation in 110-125 rad fractions delivered twice daily, 5 days per week by a ^{60}Co teletherapy unit. Two week treatment courses were alternated with two week rest periods for a total of three treatment courses. Local control was obtained in all three patients although two expired from systemic disease.

REFERENCES

1. Harrison DFN: Problems in surgical management of neoplasms arising in the paranasal sinuses. J. Laryngol. 90:69, 1976.
2. St. Pierre S, Baker SR: Squamous cell carcinoma of the maxillary sinus: analysis of 66 cases. Head and Neck Surg. 5:508-513, 1983.
3. Glick JH, Zehngebot LM, Taylor SG IV: Chemotherapy for squamous cell carcinoma of the head and neck: a progress report. Am. J. Otolaryngol 1:306-323, 1980.
4. Caradonna R, Paladine W, Ruckeschel JC, Goldstein JC, Olson JE, Jaski JW, Silvers SA, Hillinger S, Horton J: Methotrexate, bleomycin, and high-dose cis-dichlorodiam-mineplatinum (II) in the treatment of advanced epidermoid carcinoma of the head and neck. Cancer Treat Rep. 63:489-491, 1979.
5. Vogl SE, Kaplan BH: Chemotherapy of advanced head and neck cancer with methotrexate, bleomycin and cis-diamminedichloroplatinum (II) in an effective outpatient schedule. Cancer 44:26-31, 1979.
6. Sato Y, Morita M, Takahashi H, Watanabe N, Kirikae I: Combined surgery, radiotherapy and regional chemotherapy in carcinoma of the paranasal sinuses. Cancer 25:571-579, 1970.
7. Cruz AB Jr, McLnnis WD, Aust JB: Triple drug intraarterial infusion combined with x-ray therapy and surgery for head and neck cancer. Am. J. Surg. 128:573-579, 1974.
8. Zielke-Temme BC, Stevens KR Jr, Everts EC, Moseley HS, Ireland KM: Combined intraarterial chemotheapy, radiation therapy and surgery for advanced squamous cell carcinoma of the head and neck cancer. Cancer 45:1527-1532, 1980.
9. Oberfield RA, Cady B, Booth JC: Regional arterial chemotherapy for advanced carcinoma of the head and neck. Cancer 32:82-88, 1973.
10. Nervi C, Arcangeli G, Casale C, Cortese M, Guadangni A, LePera V: A reappraisal of intraarterial chemotherapy. Cancer 26:577-582, 1970.
11. Shibuyatt et al.: Reappraisal of trimodal combination therapy for maxillary sinus carcinoma. Cancer 50:2790-2794, 1982.
12. Parsons JT, Cassisi NJ, Million RR: Results of twice-a-day irradiation of squamous cell carcinoma of the head and neck. Int. J. Radiat. Oncol. Biol. Phys. 10:2041-2051, 1984.
13. Wang CC: Twice daily radiation therapy for carcinoma of the head and neck (Abstr.). Int. J. Radiat. Oncol. Biol. Phys. 7:1261-1262, 1981.
14. Jaupolis S, Pipard G, Horiot JC, Bolla M, LeDorza C: Preliminary results using twice-a-day fractionation in the radiotherapeutic management of advanced cancers of the head and neck. Am. J. Roentgenol
15. Shukovsky ML, Fletcher GH: Time-dose and tumor volume relationships in the irradiation of squamous cell carcinoma of the tonsillar fossa. Radiology 107:621-626, 1973.
16. Withers HR: Response of tissue to multiple small dose fractions. Radiat. Res. 71:24, 1977.
17. Shukovsky J, Fletcher GH, Montaque ED, Withers MD: Experience with

twice-a-day fractionation in clinical radiotherapy. Vol 126, No 1 1976.
18. Heidelberger C, Greisbach L, Montag BJ, Mooren D, Cruz D, Schnitzer RJ, Grunberg E: Studies in fluorinated pyrimidines II. Effects on transplanted tumor. Ca Res. **18**:305-317, 1958.
19. Vermund H, Hodgett J, Ansfield FJG: Effects of combined roentgen irradiation and chemotherapy on transplanted tumors in mice. AJR **85**:559-567, 1961.
20. Vietti J, Eggerding F, Valeriote F: Combined effect of x-radiation and 5 fluorouracil on survival of transplanted leukemic cells. J. Nat'l Ca Inst. **47**:865-870, 1971.
21. Douple EB, Richmond RC: Platinum complexes as radio-sensitizers of hypoxic mammalian cells. Br. J. Cancer 37, Suppl III, 98-102, 1978.
22. Douple EB, Richmond RC: A review of platinum complex biochemistry suggests a rationale for combined platinum-radiotherapy. Int. J. Radiat. Biol. Vol 5:1335-1339, 1979.
23. Salem PH et al.: Clinical phase I-II study of cis-dichlorodiammineplatinum (II) given by continuous I.V. infusion. Cancer Treatment Reports Vol. 62, No. 10, October 1978.
24. Creagen ET, Ingle JN, Schutt AJ, O'Fallon JR: A phase II study of cis-diamminedichloroplatinum and 5-fluorouracil in advanced upper aerodigestive neoplasms. Head and Neck Surgery 6:1020-1023, 1984.

HEPATIC ARTERY INFUSION (HAI) FOR HEPATIC METASTASES IN COMBINATION WITH

HEPATIC RESECTION OR HEPATIC RADIATION

H.W. Merrick*, R.R. Dobelbower, Jr.¶, J.F. Ringleint, and R.T. Skeel§
*Department of Surgery, ¶Division of Radiation Therapy
†Department of Nursing, §Department of Medicine, Medical College of Ohio at Toledo

Hepatic metastasis is the major cause of death in advanced cancer of the colon and rectum. Survival after the diagnosis of liver metastases ranges from 3-12 months (4,6,7,14). Various modes of therapy have been attempted with only partial success. Resection of isolated hepatic metastases has yielded a 15-30% five year survival (1,3,12,13). Intraarterial infusion chemotherapy using an external catheter has been performed by several investigators with response rates of 50-60% (2,8,15,16). Survival, however, averages only 8½ to 10 months, which is not significantly different from historical controls. The major problem with this therapy appears to be technical difficulties related to the external pump and catheter and the resultant necessity to terminate the intra-arterial chemotherapy (10). Radiation of the liver as an adjunct to hepatic infusion therapy has been utilized (5).

Renewed interest in hepatic artery infusion has been stimulated by the development of a totally implantable pump which eliminates many of the problems encountered by the external pumps and catheters. Initial experience by Buchwald and Grage (8) demonstrated the pump's feasibility in long term therapy. An extensive clinical experience by Ensminger (11) using floxuridine (FUDR) as the primary agent has shown HAI to be an effective modality for treatment of intrahepatic disease. He reported a tumor response rate of 83%, a 98% rate of decrease in serum CEA and a significant prolongation of survival. A study by Balch (4) also reports an 83% response rate with a low incidence of complications and side effects.

In a recent study using more strigent criteria for response (>50% decrease in tumor size on CT scan or radionuclide liver scan and decreased CEA lasting at least three months), Shepard (18) found a much lower response rate of 32% though the median survival (17 months) of patients from the start of infusion therapy was similar to that seen by Ensminger (19 months).

As the potential benefit of HAI would be greater if either all gross disease were removed by prior resection or, alternatively, if non-resectable disease were irradiated in conjunction with HAI, we initiated a phase I-II trial to evaluate combined modality therapy.

Those patients whose hepatic metastases were resectable were to receive "adjuvant" HAI for one year or until evidence of recurrence. Half of those whose metastases were non-resectable were selected to receive hepatic irradiation during the second monthly cycle of FUDR HAI.

Methods

The Infusaid totally implantable infusion pump was used in the treatment of 25 patients. The disease was staged in all patients prior to pump implantation with a hemogram and liver function tests, chest films and lung tomograms as indicated, bone scan, CT of abdomen and pelvis and superior mesenteric and celiac arteriogram. Selection criteria included good risk patients with proven hepatic metastases who had an estimated minimum survival of 2 months. Patients with proven extrahepatic metastases were excluded if this disease was extensive, could not be controlled by surgical or radiotherapy regimens or was felt to be more significant than the liver metastases in terms of survival. The arteriogram was important to demonstrate the feasibility of completely infusing the liver with the use of one or two pumps.

Twenty-five patients (13 male, 12 female) underwent implantation; nine patients had partial hepatic resection to remove all clinically evident disease in the liver and, as well, had a pump implanted for adjuvant chemotherapy to the liver. An additional seven patients who had disease throughout the liver underwent post-operative hepatic irradiation and three patients received radiation to adjacent involved lymph nodes. All patients receiving hepatic irradiation were treated with opposed anterior and posterior fields of 6 meVp or 10 meVp photons at 100 cm. focus-skin distance. All fields were individually shaped with Cerrobend alloy to conform to the liver shape as determined by clinical examination, $99^{m}Tc$ liver scan, ultrasound, CT scan and simulation. Each field was treated 5 times weekly, delivering a total dose of 2100 cGy or 2250 cGy in 150 cGy increments. Radiation was begun during the second course of FUDR hepatic infusion therapy.

Patients were initially treated with FUDR 0.3 mg/kg/day for two weeks. The drug was then unloaded from the pump and a saline-heparin solution was infused for two weeks. The cycle was then repeated with dose modification or a delay depending on the liver function tests or local/regional symptoms. When marked liver function abnormalities were seen in all patients treated at this dose for several cycles, subsequent patients were begun at a dose of 0.15 mg/kg/day. Patients were monitored with hemograms, liver function tests, perfusion scans through the pump and CT scans. Drug dose was adjusted according to toxicity.

Results

Nine patients underwent hepatic resection consisting of eight large wedge resections and one left lateral segmentectomy. Single pumps were implanted in twenty-three patients in the gastroduodenal artery, one in the splenic artery. One patient required two pumps, one in a replaced right hepatic artery and the second in the left gastric artery. There was no significant post-operative morbidity or mortality. Post-operative hepatic irradiation was carried out in 7 patients. This produced mild transient elevation of liver function tests. There have been two technical problems with the pump during the chemotherapy. One patient developed a blocked hepatic artery catheter which was cleared by streptokinase infusion and later developed a pump pocket hematoma due to an apparent spontaneous perforation of the catheter which was repaired. No other technical problems have occurred.

The response rate as determined by CT scan in 15 evaluable patients (excluding 9 undergoing resection) showed 2 (13%) complete responses (disappearance of disease), 6 (40%) partial responses (50% decrease in measured disease) and 7 (46%) non-responders, giving an overall response rate of 53%.

The response rate as measured by CEA in 20 evaluable patients (including those undergoing liver resection) showed 9 (45%) complete responders (CEA returning to normal), 6 (30%) partial responders (50% decrease in values) and 5 (25%) non responders, giving an overall response rate of 75%.

Median survival for the group of 25 patients was 10 months (1-33): For the 9 patients receiving HAI only, 8 months (2-22); for the 7 patients receiving HAI and hepatic RT, 7 months (3-26); and for the 9 patients receiving hepatic resection and HAI, 19 months (4-32). There are several long-term survivors in this group (32+ 29+ 19+ months).

Nine patients developed gastritis symptoms without major problems. Eleven patients developed one or more episodes of chemical hepatitis with jaundice. Four long-time survivors developed chronic jaundice without progressive hepatic cancer. One patient developed obstructive jaundice due to common hepatic duct necrosis which required a T-tube for long-term drainage and subsequently died of liver failure.

Ten of nineteen patients at risk survived 1 year or longer. Of these 10 patients, 7 have developed pulmonary metastasis (median interval from HAI, 14 months) 5 progressive liver disease, 1 chest wall recurrence, 1 adrenal metastasis; and 1 pump pocket recurrence.

Discussion

This study was undertaken because it was felt that the use of a combination of treatment methods might have the greatest chance of success in treating liver metastases from primary carcinoma of the colon. The project was designed as a pilot study to determine the practicality of the implantation technique, tolerance to the hepatic chemotherapy and safety of the concomitant radiotherapy.

The number of patients is too small and follow-up too short to draw any conclusions regarding the relative effectiveness of the treatment regimens. In several of the patients there has been a marked response to infusion as determined by decrease in tumor size on CT scan. Preliminary results indicate that the infusion pump is a reliable instrument, installation of which is technically feasible and infusion of chemotherapy reliable. Home care is minimal for these patients. Patient education is directed towards self-care in monitoring for the signs and symptoms of hepatitis, gastritis and elevation of temperatures which would increase the flow rate of the pump. The systemic effects of chemotherapy are minimal with no occurrence of bone marrow depression or systemic infection.

The initial concentration of FUDR of 0.3 mg/kg/day was not well tolerated and the starting dose currently employed is 0.15 mg/kg/day. This dose appears to be well tolerated with decreased incidence of hepatitis and gastritis. Careful monitoring of the liver function tests and removal of the FUDR when they were elevated has helped reduce the incidence of hepatitis. We do not know whether this decreased dose will diminish the response to chemotherapy. The surgical resections were well tolerated and in two patients who are now 31 months and 24 months post-implantation after hepatic resection, the chemotherapy has been stopped with no evidence of recurrent disease. The hepatic irradiation concurrent with hepatic FUDR infusion has been well tolerated with no evidence of long-term sequelae, and no enhancement of clinical or biochemical toxicity. Shorter survival of the irradiated patients is probably due to selection of patients with residual gross hepatic disease to receive irradiation.

In summary, the results from our study indicate that this procedure should remain investigational until its full potential and optimal utilization is determined.

REFERENCES

1. Adson, M.A. and Van Heerden, J.A.: Major Hepatic Resections for Metastatic Colorectal Cancer, Ann. Surg. 191:567-583, 1980.
2. Ansfield, F.J., Ramirez, G., Davis, H.L., Jr., et al: Further clinical studies with intrahepatic arterial infusion with 5-fluorouracil. Cancer 36:2413-7, 1975.
3. Attiyeh, F.F., Wanebo, H.J. and Stears: Hepatic Resection for Metastasis from Colorectal Cancer, Dis. Col. and Rectum, 21:160-162, 1978.
4. Balch, C.M., Urist,M.M., McGrego,M.L.: Continuous Regional Chemotherapy for Metastatic Colorectal Cancer using a Totally Implantable Infusion Pump. Am. J. of Surg., Vol. 145, 285-90, Feb. 1983.
5. Barone, R.M., Byfield,J.E., Goldfarb,P.B. et al: Intra-arterial chemotherapy using an implantable infusion pump and liver irradiation for the treatment of hepatic metastases. Cancer 50:850-62, 1982.
6. Bengmark, S., Hafstrom,L.: The natural history of primary and secondary malignant tumors of the liver. I The prognosis for patients with hepatic metastases from colonic and rectal carcinoma at laparotomy. Cancer 23:198-202, 1966.
7. Bengisson, G., Carlsson,G., Haffstrom,L., Jonsson,P.E.: Natural history of patients with untreated liver metastases from colorectal cancer. Am. J. Surg. 141:586-9, 1981.
8. Buchwald, H., Grage, T.B., Vassilopoulos, P.P., Rhode,T.D., Varco,R.L., Blackshear,P.J.: Intra-arterial infusion chemotherapy for hepatic carcinoma using a totally implantable infusion pump. Cancer 45:866-9, 1980.
9. Cady, B., Oberfield,R.A.: Regional infusion chemotherapy of hepatic metastases from carcinoma of the colon. Am J. Surg. 127:220-7, 1974.
10. Cady, B: Hepatic arterial patency and complications after catheterization for infusion chemotherapy. Ann Surg. 178:156-61, 1973.
11. Ensminger, W., Neiderhuber,J., Dakhl, S., Thrall, J., Wheeler, R.: Totally implanted drug delivery system for hepatic arterial chemotherapy. Cancer Treat. Rep. 65:383-400, 1981.
12. Fortner,J.C., et al: Major Hepatic Resection for Neoplasm: Personal Experience in 108 patients, Ann. Surg. 188:363-370, 1978.
13. Foster, J.H.: Survival After Liver Resection for Secondary Tumors, Am J. Surg. 135:389-394, 1978.
14. Goslin, R., Steele, G., Zamchjid, N., Mayer, R., MacAntyre, J.: Factors influencing survival in patients with hepatic metastases from adenocarcinoma of the colon or rectum. Disease of Colon & Rectum, Vol. 25, No. 8, 749-753, Nov.-Dec., 1982.
15. Jaffe, B.M., Donegan, W.L., Watson, F., Spratt, J.S.: Factors influencing survival in patients with untreated hepatic metastases. Surg. Gynecol. Obstet 127:1-11, 1968.
16. Oberfield, R.A., McCaffrey, J.A., Polio, J., Clouse, M.E., Hamilton, T. Prolonged and continuous percutaneous intra-arterial hepatic infusion chemotherapy in advanced metastatic liver adenocarcinoma from colorectal primary. Cancer 44:414-23, 1979.
17. Sullivan, R.D.: Systemic and arterial infusion chemotherapy for metastatic liver cancer. Int. J. Radiat. Oncol. Biol. Phys. 1:973-6, 1976.
18. Shepard,K.V., Levin,B., Karl,R.C., Faintuch,J., DuBrow,R.A., Hagle,M., Cooper,R.M., Beschorner,J., and Stablein,D.: Therapy for metastatic

colorectal cancer with hepatic artery infusion chemotherapy using a subcutaneous implanted pump, J. Clin. Oncol. 3, 161-169, 1985.

PHASE II STUDY OF SIMULTANEOUS RADIATION THERAPY

CONTINUOUS INFUSION 5-FU AND BOLUS MITOMYCIN-C

        Oscar A. Mendiondo, Lawerence C. Maguire,
        William D. Medina, and John D. Cronin

        Lexington Clinic Cancer Center
        1221 S. Broadway
        Lexington, KY 40504

SUMMARY

The combination of radiation therapy with a continuous infusion of 5-Fluorouracil and bolus injection of Mitomycin-C has been successful in the treatment of carcinomas of the anal canal and also squamous-cell carcinomas of the female genitalia and of the head and neck. One hundred and ten patients with a variety of carcinomas selected because of their expectedly poor response to standard treatment have been treated at our institution with a combination of radiation and chemotherapy. Best results were obtained in patients with squamous-cell carcinomas of the anal canal, esophagus, head and neck and cervix, with 70% of evaluable patients demonstrating a complete response and 21% showing a partial response. Patients with adenocarcinomas of the gastrointestinal tract faired poorly with only a 61% response, most of them partial. While this particular combination of chemotherapy and radiotherapy appears promising in squamous cell carcinomas of the head and neck, esophagus, anal canal and uterine cervix, results in other malignancies appear far less adequate and suggest that other combinations will have to be examined.

INTRODUCTION

There continue to be all too many malignancies for which effective treatment is at best limited. Multimodality therapy is at present widely used but the most appropriate combination and scheduling of surgery, radiation therapy and chemotherapy is still to be found. The use of concomitant continuous infusion chemotherapy and simultaneous radiation therapy has been found to be curative, with or without surgery, in squamous-cell carcinomas of the anal canal[1] and has also found to be promising in other tumors such as carcinomas of the esophagus[2], advanced head and neck malignancies[3] and carcinomas of the cervix[4]. In this paper, we present our preliminary experience in the treatment of 110 patients with different types of neoplasms, who have been treated with a combination of radiation, continuous infusion of 5-Fluorouracil (5-FU) and bolus Mitomycin-C (MC).

MATERIALS AND METHODS

The patients selected for this particular mode of therapy were those with carcinomas of the anal canal, esophagus and advanced carcinomas of the head and neck, cervix, pancreas, rectum, stomach, lung or bladder.

Table I. Combined 5-FU, Mitomycin-C and Radiotherapy
Sites Treated and Evaluability

Site	No. Pts. Treated	No. Pts. Evaluable
Anal Canal	9	8
Esophagus	16	15
Head & Neck	19	17
Cervix	3	3
Pancreas	11	6
Rectum	25	15
Stomach	6	5
Lung	14	9
Bladder	7	6
TOTAL	110	84

Performance status greater than 30% on the Karnofsky performance scale was required. Patients with metastatic disease or prior radiotherapy to the involved area were accepted as long as they were evaluable for their local response. Table I shows the number of patients treated and the number of patients evaluable for each disease category. Inevaluable were those patients with a follow up too short to permit adequate assessment of response.

Each cycle of chemo- radiotherapy consisted of three weeks of radiation, given in successive weeks for thoracic, abdominal and pelvic malignancies and with a one week interruption for head and neck malignancies. A continuous infusion of 5-FU (20 mg/kg/day) was given with the first and third weeks of treatment and a single bolus injection of MC (10 mg/m^2) was given the first day of treatment. Radiation therapy was given to doses of 1000 rad per week, with a total dose of 3000 rad in three weeks (four weeks for head and neck tumors). Complete response was defined as absence of disease in the treated area, either by a negative surgical specimen in patients who underwent surgery or in all others by clinical examination, repeat endoscopy with biopsies, diagnostic x-rays or CAT scans. Partial response was estimated as a 50% or better reduction in evaluable disease parameters.

Results

Table II shows the results obtained in eight evaluable patients with cancer of the anal canal. Only two patients received two cycles of chemo-radiotherapy while the other six were treated with only one cycle. Seven of the eight patients had a complete response with a duration of 4 to 41 months, with the only patient who had a partial response being converted to a complete response by surgery. With the exception of one patient

TABLE II. Combined 5-FU, Mitomycin-C and Radiotherapy
Cancer of Anal Canal (8 evaluable patients)

Patient #	Histology	TNM Stage	Response	Follow Up (mos.)
1	Cloacogenic	T_4 N_1 M_0	Complete	41
2	Cloacogenic	T_3 N_0 M_0	Complete	40
3	Squamous	T_{2b} N_0 M_0	Complete	18
4	Cloacogenic	T_{2b} N_0 M_0	Complete	16
5	Squamous	T_3 N_0 M_0	Partial	15
6	Squamous	T_3 N_0 M_0	Complete	14
7	Squamous	T_4 N_0 M_0	Complete	10
8	Basaloid	T_{2b} N_0 M_0	Complete	4

TABLE III. Combined 5-FU, Mitomycin and Radiotherapy
Head and Neck Cancer

Site Treated	No. Patients	Complete Response	Follow Up (mos.)
Larynx	4*	3	(3), 5, 8, 14
Base of Tongue	3*	2	(0), 19, 41
Posterior Pharynx	2	2	4, 8
Nasopharynx	1	1	15
Cervical Esophagus	1	1	13
Tonsil	1	1	6
Gingiva	1**	0	(15)
Neck Nodes (T_{0-X})	3	1	(4), (6), 9
Parastomal Recurrence	1	1	6

* one patient did not complete treatment
** preoperative treatment
() follow up of patients with partial or no response

who died of lung cancer four months after treatment of his anal carcinoma, the other patients are at present alive and free of metastases.

Of the 15 evaluable patients with esophageal cancer, nine had a complete response and five had a partial response. Five of the 15 patients underwent surgery after preoperative chemo- radiotherapy, with three of them having a negative specimen. Esophagitis during therapy was almost a rule in all these patients but did not compromise their status.

Eleven of the 17 evaluable patients with head and neck malignancies had advanced tumors (T_{3-4}, N_{2-3}). The five patients with earlier lesions (T_{1-2}, N_{0-1}) had lesions of the base of the tongue, posterior pharyngeal wall or hypopharynx. One patient had a parastomal recurrence after surgery and radiation for laryngeal cancer. The results obtained in these patients are shown in Table III. Twelve of 17 evaluable patients had a complete response. Of the patients who had only a partial response or no response, one refused a second cycle of therapy, one was treated preoperatively, having residual disease in the surgical specimen, two were palliatively treated for massive neck disease and one developed a severe mucositis which lead to important malnutrition and discontinuation of therapy. Only one of the patients who entered a complete response has so far recurred. This patient had an advanced

TABLE IV. Combined 5-FU, MC and Radiotherapy
Response by Site
(84 evaluable patients)

Site	Evaluable	CR	PR	NR
Anus	8	7	1	-
Esophagus	15	9	5	1
Head & Neck	17	12	2	3
Cervix	3	2	1	-
Subtotal	(43)	(30)	(9)	(4)
Pancreas	6	1	4	1
Rectum	15	1	6	8
Stomach	5	1	3	1
Lung	9	2	3	4
Bladder	6	1	3	1

CR: complete response
PR: partial response
NR: no response

carcinoma of the larynx and was treated with a preoperative intent, then refusing further treatment.

Table IV shows a breakdown of response by disease site. Two of three patients with carcinoma of the cervix had a complete response. The one partial response was on the basis of microscopic residual disease in a hysterectomy specimen in a patient treated preoperatively. Overall, good responses were observed in carcinomas of the anus, esophagus, head and neck and cervix, with 30 complete responses in 43 patients (69.8%). Response was poor in carcinomas of the pancreas (1/6), rectal adenocarcinomas (1/15), stomach (1/5), lung (2/9), and bladder (1/6).

Treatment related mortality was seen in two patients. One of them, with a cancer of the base of the tongue and who had received chemotherapy previously, developed severe mucositis after the first week of treatment necessitating discontinuation of therapy; he later developed sepsis and expired. One patient with squamous-cell carcinoma of the esophagus treated preoperatively, developed a myocardial infarction 30 days after surgery. Severe morbidity was seen in three patients. Two of them had advanced squamous-cell carcinoma of the esophagus and developed a tracheoesophageal fistula requiring feeding jejunostomies. One patient with squamous-cell cancer of the cervix who was treated preoperatively, developed recurrent tumor and a rectovesicovaginal fistula. Severe mucositis in head and neck patients was averted by the use of a week's rest after the first week of therapy in each cycle. Important weight loss was only seen once. Treatment of thoracic, abdominal and pelvic malignancies was uniformly well tolerated.

DISCUSSION

The combination of chemotherapy and radiation therapy for the treatment of human malignancies is theoretically attractive since both modalities interact in a variety of ways to achieve tumor cell eradication. This is, of course, of more interest in malignancies where radiation alone has traditionally been of limited success. Ever since Nigro[5] reported excellent response of tumors of the anal canal to a combination of radiation therapy with infusional 5-FU, there has been an increasing interest in this particular treatment combination. Other investigators[1,6], have confirmed that experience and have extended the application of similar regimens to the treatment of malignancies of the head and neck[3], esophagus[2] and cervical cancer[4].

Our experience parallels that of other groups, with excellent initial results being seen in carcinomas of the anal canal and squamous-cell cancers of the esophagus, head and neck and cervix. The regimen we employed differs from the one initially reported by Nigro[5] in that only one dose of MC is given with the first day of treatment and the total duration of each cycle is shortened to three weeks by the incorporation of the second 5-FU infusion into the third week of radiation. This regimen has been almost uniformly well tolerated. The interruption of therapy for one week after the first week of radiation therapy and 5-FU for head and neck malignancies, has also been well tolerated even in patients requiring large field irradiation encompassing oropharynx and hypopharynx.

Most of our patients with carcinomas of the anal canal have received only one cycle of chemo- radiotherapy and have demonstrated persistent excellent local control. This raises doubts about the need for any higher doses. Accordingly, our present practice is to simply repeat deep biopsies after that first course of chemo- radiotherapy and offer surgery to those patients who have not been controlled, which seems to be a rather infrequent event.

The results with esophageal cancer pose questions of a different nature. It appears that combined modality therapy can offer a reasonable likelihood of local control and future trials should possibly address the role of surgery. The initial results obtained in carcinomas of the head and neck are at least exciting. With most of our patients having fairly advanced tumors, complete responses after the first cycle of chemo- radiotherapy have been the rule rather than the exception. Although our follow-ups are short, findings of other authors[3] suggest that 70% of these responses will be long lasting. It is interesting that one of our patients, who was treated palliatively because of prior high dose irradiation to her oropharynx, has shown a persistent complete response after only one cycle of chemo- radiotherapy suggesting that at least some patients might not need what is considered a standard dose of irradiation for control of their tumors. Although only three patients with carcinomas of the cervix have been treated, encouraging results have also been seen in these advanced tumors.

Our results with carcinomas of the lung with a variety of non-small cell types, adenocarcinomas of the pancreas, rectum and stomach and transitional-cell carcinomas of the bladder, have been rather disappointing. Although stabilization of disease has been achieved in a number of patients, complete responses seem to be anecdotal.

Although follow-ups are too short, cautious optimism is quite justified in view of the initial results obtained with this particular combination of radiation therapy, 5-FU and MC. Further studies should be undertaken in an attempt to determine total doses necessary, best interval between cycles and best means for the assessment of a local response. Careful follow up will be necessary to assess longterm effects of combined therapy on normal tissues. Other combinations or different schedules will have to be considered for those tumors that appear to be poorly responsive to this combination of drugs and radiation.

REFERENCES

1. B. Sischy, J. H. Remington, J. Hinson, S. H. Sobel, J. E. Woll, Definitive treatment of anal canal carcinoma by means of radiation therapy and chemotherapy, Dis. Colon Rectum 25:685 (1982).
2. Z. Steiger, R. Franklin, R. F. Wilson, L. Leichman, H. Seydel, J. J. K. Loh, G. Vaishamapayan, T. Knechtges, I. Asfaw, A. Dindogru, J. C. Rosenberg, T. Buroker, A. Torres, D. Hoschner, P. Miller, T. Pietruk, V. Vaitkevicius, Eradication and palliation of squamous cell carcinoma of the esophagus with chemotherapy, radiotherapy and surgical therapy, J. Thorac. Cardiovasc. Surg. 82:713 (1981).
3. J. E. Byfield, T. R. Sharp, S. S. Frankel, S. G. Tang, F. B. Callipari, Phase I and II trial of five-day infused 5-Fluorouracil and radiation in advanced cancer of the head and neck, J. Clin. Oncol., 2:406 (1984).
4. G. Thomas, A. Dembo, F. Beale, H. Bean, R. Bush, J. Herman, J. Pringle, G. Rawlings, J. Sturgeon, S. Fine, B. Black, Concurrent radiation Mitomycin-C and 5-Fluorouracil in poor prognosis carcinoma of the cervix: preliminary results of a phase I-II study, Int. J. Rad. Oncol. Biol. Phys. 10:1785 (1984).
5. N. D. Nigro, G. K. Vaitkevicius, G. Considine, Combined therapy for cancer of the anal canal: a preliminary report, Dis. Colon Rectum 17:354 (1984).
6. B. Cummings, T. Keane, G. Thomas, A. Harwood, W. Rider, Results and toxicity of the treatment of anal canal carcinoma by radiation therapy or radiation therapy and chemotherapy, Cancer 54:2062 (1984).

CANCER OF THE ESOPHAGUS —
MEDICAL UNIVERSITY OF SOUTH CAROLINA
PILOT STUDY

J. M. Jenrette III, R. D. Marks Jr.,
E. F. Parker, and R. H. Fitzgerald Jr.

Medical University of South Carolina

171 Ashley Avenue
Charleston, South Carolina 29425

INTRODUCTION

Cancer of the esophagus is a rare neoplasm accounting for 2% of all cancer deaths each year.[1] The survival remains poor despite improved staging, post-operative care, and adjuvant therapy. Our previous experience is summarized in Table 1. During the period 1940-1951, the majority of the patients were treated surgically when considered operable. The two-year survival remained consistently less than 1% during the decade. Throughout the next twenty-six years, radiation therapy was added to the treatment plan, first as a definitive treatment and later as an adjuvant to therapy. Cures with any modality remained uncommon.

In 1980, based on our own previous dismal survival, we embarked upon a combined approach of preoperative chemotherapy and radiation therapy. Several centers had engaged similar protocols at this time and subsequently published their data.[2,3,4]

METHODS AND MATERIALS

During the period 1980-1984, 129 patients with a diagnosis of carcinoma of the esophagus were seen at the Medical University of South Carolina. Eighty-nine were placed on protocol, and 40 patients were not. Those who were not placed on protocol were seen by members of the Thoracic Surgery Department who were not participants in the pilot study. The breakdown of patients by race, sex and age may be found in Table 2. All patients who entered the study were staged with endoscopy and biopsy. Barium swallow, upper GI, and chest x-rays were done on all patients. A CAT scan was performed to assess resectability. A complete blood count, liver function tests, and routine chemistries were obtained. Bilateral supraclavicular fossae node biopsies were performed on all patients. A positive finding would have precluded continuation in the study. The study objectives were to investigate combined chemotherapy and radiation therapy with esophageal resection on local control and survival, to extensively stage patients, and to determine pathologically the effects of combined chemotherapy and radiation therapy on the primary tumor.

The radiation therapy phase consisted of 3000 rad tumor dose given in 15 fractions over three weeks. A Cobalt-60, 4 MeV, or 10 MeV therapy unit was used in each case. Port arrangements were designed to include the tumor with a 6 cm margin superiorly and inferiorly, whenever possible.

The chemotherapy phase began on Day 1 of the radiation therapy. The drugs administered were 5-Fluouracil (1000 mg pen M^2 in 100 cc of dextrose and water per day for four days) and Mitomycin-C 10 mg. IV push on day 1.

Following the course of chemo-radiation therapy, the patient was restaged for esophagectomy.

TABLE I

CARCINOMA OF ESOPHAGUS
PRIOR EXPERIENCE

SERIES	CASES	OPERABLE	RESECTABLE	MORTALITY	TWO-YEAR SURVIVORS	
1940-51	170	56 (33%)	30 (54%)	17 (57%)	1	(0.6%)
1951-57	166	39 (23%)	34 (87%)	19 (56%)	9	(5.0%)
1957-62	135	30 (22%)	21 (70%)	3 (14%)	11	(9.0%)
1962-67	138	47 (34%)	41 (87%)	13 (31%)	17	(12.0%)
1967-75	300	116 (39%)	90 (80%)	18 (20%)	23	(7.7%)

TABLE 2

CARCINOMA OF THE ESOPHAGUS
MAY 1980 - APRIL 1984

BLACK MALE	88	59.5 YRS.	31-81 YRS.
WHITE MALE	25	59.2 YRS.	49-80 YRS.
BLACK FEMALE	12	66.8 YRS.	46-78 YRS.
WHITE FEMALE	4	69.3 YRS.	63-79 YRS.
TOTAL	129	60.4 YRS.	31-81 YRS.

RESULTS

Of the 129 patients with cancer of the esophagus during this period, we have 98 patients with two year follow-up. (Table 3) Of significance are the 21 patients in group 6 who received radiation therapy, chemotherapy and resection. Of those 21 patients, 7 are two year survivors. Of this select group survival exceeds previous results from this institution.

Additionally, the percent of positivity in the resections was 68% which compares favorably with previous findings of 87% in earlier series. (Table 4)

DISCUSSION

The regimen was well tolerated in our patients with a performance status satisfactory to undergo the regimen. The mortality rate was reduced from 25% in a previous series to 9.7% in the present one. Although there is no clear survival advantage at present, we feel that the decrease in positive resected specimens may be an important factor.

In future trials, we will increase the dose of radiation to 4500 rad. We will conduct a prospective trial testing patients for surgery above, radiation therapy alone, preoperative radiation therapy, and preoperative chemo-radiation therapy. We will look for factors to define a subset of patients whose tumors have been sterilized pathologically who do not need resection.

TABLE 3

CARCINOMA OF THE ESOPHAGUS (MAY 1980 - OCT 1982 N=98)	NON PROTOCOL	2 YR SURVIVORS	PROTOCOL	2 YR SURVIVORS
1) NO TREATMENT	3	0	9	0
2) MRT++ ALONE	21*	1	12	0
3) CHEMOTHERAPY ALONE	0		2	0
4) MRT & CHEMOTHERAPY x 1	0		10*	1+
5) MRT & CHEMOTHERAPY x 2	0		3	0
6) MRT + CHEMO + RESECTION	0		21	7(33%)**
7) NO PREOP RX - OP-NON-RESECTION	1	0	1	0
8) NO PREOP RX - RESECTION	5	2	0	
9) NO PREOP RX - RESECT-POSTOP MRT	1	0		
10) PALL. BY-PASS ALONE	1	0	3	0
11) PALL. BY-PASS + POSTOP MRT	0		5	0
TOTAL	32	3(9.4%)	66	8(12.1%)

*ONE PATIENT OPERABLE BUT REFUSED OPERATION
+PATIENT HAS METASTATIC TUMOR
**ONE PATIENT HAS METASTATIC TUMOR
++MEGAVOLTAGE RADIATION THERAPY

TABLE 4

CARCINOMA OF THE ESOPHAGUS

RESIDUAL TUMOR IN RESECTED SPECIMENS
AFTER PRE-OP CHEMO-RADIOTHERAPY

	RESECTIONS	POSITIVITY
1980-81	8	5
1981-82	8	5
1982-83	9	6
1983-84	6	5
TOTALS	31	21 (68%)

AMONG 3 YR. SURVIVORS 1 POS 1 NEG
AMONG 2 YR. SURVIVORS 1 POS 1 NEG
AMONG 1 YR. SURVIVORS 2 POS 2 NEG

REFERENCES

1. Silverberg, E. Cancer Statistics, 1985, CA - A Can. J. for Clin., 35:26-27 (1985)
2. Steiger, Z., Franklin, R., Wilson, R.F., Leighman, L., Segdel, H., Loh, J.F.H. Eradication and Palliation of Squamous Cell Carcinoma of the Esophagus with Chemotherapy, Radiotherapy and Surgical Therapy, J. Thor and C-V Surg., 2:713 (1981).
3. Bains, M.S., Kelson, D.P., Beattie, E.J. and Marlin, N. Treatment of Esophageal Carcinoma by Combined Preoperative Chemotherapy, Ann. Thor. Surg., 34:521 (1982).
4. Kelsen, D.P., Bains, M., Chapman, R. and Tolbey, R. Cisplatin, Vindesine, and Bleomycin PVB Combination Chemotherapy for Esophageal Carcinoma, Can. Treat. Rep., 65:781 (1981).

CONTINUOUS INFUSION VP-16, BOLUS CISPLATIN, AND
SIMULTANEOUS RADIATION THERAPY AS SALVAGE THERAPY
IN SMALL CELL BRONCHOGENIC CARCINOMA

K. Rowland Jr., P. Bonomi, S. Taylor, S. Maffey,
S. Reddy, and M.S. Lee

Rush University, Section of Medical Oncology
Chicago, IL

INTRODUCTION

VP-16 is regarded as one of the most active single agents in small cell bronchogenic carcinoma (SCBC). As second line therapy response rates have ranged from 0-40% with a cumulative series response rate of 12% in 381 patients.[1-10] The importance of VP-16 scheduling is suggested by in vitro data revealing enhanced cytotoxicity after prolonged exposure and clinical data showing the benefit of frequent, daily schedules over weekly administration.[11,12] Cisplatin as second line therapy for SCBC has demonstrated significant activity in 8 of 54 patients.[13,14,15]

Together cisplatin and VP-16 appear to have synergistic cytotoxicity in a variety of malignancies particularly SCBC in which response rates of 44-55% have been reported when this regimen was used as salvage therapy.[16-22] Radiation therapy alone has been reported to control SCBC failing after initial chemotherapy.[23] The ability of cisplatin to potentiate radiation has been observed in in vitro and in vivo studies.[24,25] There has been an anecdotal report of podyphyllins acting as a radiation sensitizer as well.[26] Continuous infusion VP-16, bolus cisplatin, and simultaneous, hyperfractionated radiation have been combined in an effort to maximize their potential synergistic interactions.

PATIENTS AND METHODS

Patients with histologically confirmed SCBC refractory to previous therapy were evaluated. Eligibility requirements included histologically or cytologically confirmed SCBC, measurable disease, no previous radiation therapy, failure or progression on one or more chemotherapy regimens, performance status 3 or better, and adequate hematological, renal and biliary function (WBC > $4000/mm^3$, platelet count > $100,000/mm^3$, BUN < 25 mg/dl, creatinine < 1.5 mg/dl, and bilirubin < 2.0 mg/dl).

Concomitant daily or twice daily radiation therapy was given day 1 through day 5 along with continuous infusion VP-16 (see figure 1). Total infusion time was 96 hours beginning shortly after radiation therapy on day 1 and ending just after radiation therapy on day 5. Cisplatin was given over one hour with saline and mannitol diuresis on day one. Each 5 day cycle was scheduled every 2 to 3 weeks. All patients began at 60 mg/m^2 per day of VP-16 and 60 mg/m^2 of cisplatin, except those with performance status 3 for which initial doses were 50% lower. Dose modifications were made on the basis of day one hematologic and renal values. Involved field radiation therapy to symptomatic sites was administered with simultaneous chemotherapy for up to four cycles (4000 cGy). Thereafter, patients received either one or two more combined modality therapy cycles with shrinking radiation fields, maintenance chemotherapy, or observation at investigator discretion. Standard ECOG criteria were utilized to evaluate progression and response.

Schedule:

VP-16	60 mg/m^2	96 hour infusion d1-5
cisplatin	60 mg/m^2	1 hour infusion d1
RT	116 cGy	BID d1-5

FIGURE 1

RESULTS

Fifteen patients have been entered on study and all are evaluable. Patient characteristics are listed in table I. Eleven failed CAV chemotherapy and 5 had received VP-16 prior to study. Nine patients had responded to initial chemotherapy and 3 progressed while receiving their initial cycles of first line therapy. Median time from diagnosis to initiation of VP-16, cisplatin, and radiation therapy was 5.7 months.

All patients received at least 2 cycles of combined modality therapy (median 4). Of ten patients treated initially at 2 week intervals, 15 of 22 cycles were modified either by delay in treatment or by dose reduction for leukopenia or thrombocyotopenia. For 5 patients treated at 3 week intervals (total of 16 cycles), treatment was delayed once and dose was decreased in 7 because of myelosuppression. The percents (actual/full dose) of cisplatin and VP-16 were 79% and 75% respectively. Six patients were given maintenance chemotherapy for 1-5 cycles (median 2).

All but one patient whose dominant site of disease was hepatic received thoracic radiation therapy. Five patients were treated with twice daily fractions. Median dose of radiation was 4000 cGy (range 2000-5900).

Of 15 patients, six (40%) achieved a clinical complete response and five (33%) achieved a partial response. Only one patient showed progressive disease while on therapy. Six of the responding patients have failed with a median duration of response of three months. A total of 12 patients have failed with a median time to failure of 3.8 months. Actuarial survival from the time of diagnosis and of salvage therapy for all patients is 16.8 months and 5.7 months respectively.

Of the 12 patients who have failed, four failed within the radiation port only; two both inside and outside of the field, and six at distant sites only. Three of four patients at autopsy were found to have tumor within the radiation field.

Of WBC and platelet counts checked every two to three weeks throughout treatment, the median lowest WBC and platelet counts were 2.1 mm^3 (range 1.5-7.6) and 112 mm^3 (range 20-304) respectively. There were no leukopenic related infections. No patients required platelet transfusions. There were no episodes of dermatitis, mucositis, or cardiac compromise. Nausea was controllable in all patients. Alopecia was noted in nearly all patients.

TABLE 1

Patient Characteristics

Age (Median)	57
Limited Disease	10
Extensive Disease	5
Prior Chemotherapy	
One	10
Two or more	5
Performance Status	
0,1	7
2	6
3	2

DISCUSSION

Of the fifteen patients studied the overall response rate of 73% and the actuarial survival from the time of salvage of 5.7 months suggests significant activity of this combined modality regimen in relapsing SCBC. Utilizing chemotherapy alone for salvage in SCBC, response rates have usually been less than 30% with survival from the time of relapse less than 2 months.[27] Of late, higher salvage response rates have been achieved with bolus cisplatin (D) and VP-16 (V). Evans[16], Porter[18], and Tinsley[19] have independently reported response rates of 47-55%. In Tinsleys' study of 16 patients treated with DV (D -20 mg/m^2 days 1-5 and V - 100 mg/m^2 bolus days 1-5 every 3 weeks), 7 (44%) patients responded. Three patients died of treatment related complications. No survival data was reported. Of 28 patients treated by Porter with DV (D - 40 mg/m^2 days 1-3 and V - 200 mg/m^2 bolus days 1-3 every 3 weeks), 15 (54%) patients responded. Median survival for responders and all patients after salvage was 5 months and 3 months respectively. Over one-third of the patients developed severe myelosuppression. Evans reported a 55% salvage response rate with D plus bolus V. The dose and schedule initially used, D - 20 mg/m^2 days 1-5 and V -100 mg/m^2 bolus days 1, 3, and 5, were later modified to D - 25 mg/m^2 days 1-3 and V -100 mg/m^2 bolus days 1-3. Of 78 evaluable patients failing CAV, some with mediastinal radiation, median survival from the time of salvage for responders was 10.5 months. Ochs[23] has reported a 64% response rate with radiation therapy as salvage for SCBC. Of note, 14 of the 25 evaluable patients also received unspecified concurrent chemotherapy, and 11 of 16 responders subsequently progressed within the radiation port. Median survival after salvage for all 25 patients was 3.7 months.

Two reports to date have attempted to exploit the combined modality activity of D plus bolus V and radiation therapy in SCBC by scheduling simultaneous thoracic radiation with at least one cycle of DV. Murray[28] treated 67 previously untreated patients with alternating D plus bolus V and RT with CAV. Murray et al noted a 94% response rate with a median survival of 20.5 months. Woods[29] employed concomitant thoracic radiation therapy (to 3900 cGy) during the first two cycle of D plus bolus V. Overall response rate was 71% and median survival for all 148 previously untreated patients was 12.5 months. No "synergistic" toxicities were mentioned.

Scheduling VP-16 as a continous infusion as used in our study was based on in vitro observations that prolonged exposure to VP16 resulted in greater cytotoxicity.[11] However, it should be pointed out that a single randomized clinical trial showed no clear advantage for continuous infusion over bolus VP16 in SCBC.[30] Cisplatin was not included in the regimen studied in this trial.

Based on these preliminary results it appears that toxicity with comcomitant radiation therapy and D infusion V is acceptable and that within the radiation port activity is relatively high in previously treated SCBC patients. Incorporation of this regimen as initial treatment of SCBC patients is planned.

REFERENCES

1. Cohen M, Broder L, Fossieck B, et al. Phase II clinical trial of weekly administration of VP16-213 in small cell bronchogenic carcinoma. Cancer Treat. Rep. 61:489-490, 1977.

2. Hansen M, Hirsch F, Dombernowsky P, et al. Treatment of small cell anaplastic carcinoma of the lung with the oral solution of VP16-213. Cancer 40:633, 1977.

3. Ossell B, Einhorn L, Comis R, et al. Multicenter phase II trial of etoposide in refractory small cell lung cancer. Cancer Treat. Rep. 69:127-128, 1985.

4. Antman K, Pomfret E, Kays G, et al. Phase II trial of etoposide in previously treated small cell carcinoma of the lung. Cancer Treat. Rep. 68:1413-1414, 1984.

5. Dally M, Harper P, Smyth J, et al. Epipodophyllotoxin (VP-16/213) in small cell carcinoma of the bronchus resistant to initial combined chemotherapy. British J. of Diseases of the Chest 312, 1980.

6. Karp G, Antman K, Cannellos G. VP-16-213 a phase II trial using a weekly schedule. Cancer Clin. Trials 4:465-467, 1981.

7. Tempero M, Kessinger A, Lemon H. VP-16-213 therapy in patients with small cell carcinoma of the lung after failure on combined chemotherapy. Cancer Clin. Trials 4:155-157, 1981.

8. Tucker R, Ferguson A, VanUyk C, et al. Chemotherapy of small cell carcinoma of the lung with VP 16-213. Cancer 41:1710, 1970.

9. Nissen N, Rajak T, Leone L, et al. Clinical trial of VP 16-213 IV twice weekly in advanced neoplastic disease. Cancer 45:232-235, 1980.

10. Eagan R, Carr D, Frytak S, et al. VP-16-213 versus polychemotherapy in patients with advanced small cell lung cancer. Cancer Treat. Rep. 60:949-951, 1976.

11. Dombernowsky P and Nissen N. Schedule dependency of the antileukemic activity of the podophyllotoxin derivative VP16-213 in L1210 leukemia. Arch. Path. Micro. Scandinavia 81:715-724, 1981.

12. Sierocki, J, Hilaris B, Hopfan S, et al. cis-Dichlorodiamineplatinum (II) and VPP-16-213: an active induction regimen for small cell carcinoma of the lung. Cancer Treat. Rep. 63:1593, 1979.

13. Dombernowsky P, Sorenson S, Aisner J, et al. cis-Dichlorodiammineplatinum (II) in small cell anaplastic bronchogenic carcinoma: a phase II study. Cancer Treat. Rep. 63:543-545, 1979.

14. Bhuchar V, Canzotti V. High-dose cisplatin for lung cancer. Cancer Treat. Rep. 66:375-376, 1982.

15. Cavalli F, Goldhirsch A, Siegenthaler P, et al. Phase II study with cis-dichlorodiammineplatinum (II) in small cell anaplastic bronchgenic carcinoma. European J. of Cancer 16:617-621, 1980.

16. Evans W, Osoba D, Feld F, et al. Etoposide (VP-16) and cisplatin: an effective treatment for relapse in small cell lung cancer. J. Clin. Oncol. 3(1):65-71, 1985.

17. Madrigal P, Manga G, Palomero T, et al. VP-16-213 combined with cis-platinum in the treatment of small cell carcinoma of the lung. Cancer Chemo. Pharm. 7:203-204, 1982.

18. Porter L, Johnson D, Hainsworth J, et al. Recurrent small cell lung carcinoma treated with VP-16 and cisplatin. Am. Soc. Clin. Oncol. C-840, 1984.

19. Tinsley R, Comis R, DiFino S, et al. Potential clinical synergy observed in the treatment of small cell lung cancer with cisplatin and VP-16-213. Proc. Am. Soc. Clin. Oncol. C-772, 1983.

20. Kim P and McDonald D. The combination of VP-16-213 and cis-platinum in the treatment of small cell carcinoma of the lung. Proc. Am. Soc. Clin. Oncol. C-547, 1982.

21. Lopez J, Mann J, Grapski R, et al. Salvage chemotherapy of refractory small cell lung cancer with VP-16 and cis-platinum. Am. Soc. Clin. Oncol C-586, 1982.

22. Evans W, Feld R, Osoba D, et al. VP-16 alone and in combination with cisplatin in previously treated patients with small cell lung cancer. Cancer 53:1461-1466, 1984.

23. Ochs J, Tester W, Cohen M, et al. "Salvage" radiation therapy for intrathoracic small cell carcinoma of the lung progressing on combination chemotherapy. Cancer Treat. Rep. 67:1123-1126, 1983.

24. Douple E, Richmond, R. A review of platinum complex biochemistry suggests a rationale for combined platinum radiotherapy. Int. J. Rad. Onc. Biol. Phys. 5:1335-1339, 1979.

25. Fu K, Rayner P, Lam K. Modification of the effects of continuous low dose rate irradiation by concurrent chemotherapy infusion. Int. J. Rad. Onc. Biol. Phys. 10:1477-1478, 1984.

26. Swartz M, Roseman J, Gunderson L, et al. Podophyllin increases tumor radioresponsiveness. Cancer Chemo. Pharm. 7:238, 1982.

27. Morstyn G, Ihde D, Lichter A, et al. Small cell lung cancer 1973-1983: early progress and recent obstacles. Int. J. Rad. Oncol. Biol. Phys. 10:515-539, 1984.

28. Murray N, Hadzic E, Shah A, et al. Alternating chemotherapy and thoracic radiotherapy with concurrent cisplatin for limited stage small cell carcinoma of the lung. Proc. Am. Soc. Clin. Onc. C-835, 1984.

29. Woods R, Levi, J. Chemotherapy for small cell lung cancer: a randomized study of maintenance therapy with cyclophosphamide, adriamycin, and vincristine after remission induction with cisplatinum, VP-16-213, and radiotherapy. Proc. Am. Soc. Clin. Oncol. C-836, 1984.

30. Aisner J, Whitacre D, VanEcho D, et al. Doxorubicin, cyclophosphamide, and etoposide (ACE) by bolus or continuous infusion for small cell carcinoma of the lung. Proc. Am. Soc. Clin. Onc. C-765, 1983.

31. Willson J, Neville A, Davis T. A phase II trial of etoposide infusion, cis-platinum, and hexamethylemlamine in small cell carcinoma of the lung. Proc. Am. Soc. Clin. Onc. C-897, 1984.

32. Aisner J, VanEcho D, Whitacre M, et al. A phase I trial of continuous infusion VP-16-213 (etoposide). Cancer Chemo. Pharm. 7:157-160, 1982.

CONCOMITANT RADIATION, MITOMYCIN C AND 5 FLOUROURACIL INFUSION IN

GASTROINTESTINAL CANCER - A PRELIMINARY REPORT

A. Chan, A. Wong, and K. Arthur

Tom Baker Cancer Centre
1331-29 Street, N.W., Calgary
Alberta, Canada, T2N 4N2

For a decade, concomitant radiation and chemotherapy with 5-Fluorouracil and Mitomycin C has been used in treating carcinoma of anal canal with favourable results (Nigro[1], Cummings[2], Byfield[3]). Similar encouraging results have been reported with esophageal carcinoma (Byfield[4], Franklin[5], Keane[6]). To date, this combination therapy has been used chiefly in squamous cell carcinoma. It's efficacy in adenocarcinoma is not well documented, though felt to be ineffective by some authors. Wassif[7] has employed similar combination in patients with advanced stages of rectal carcinoma, and reported local response rate of 79-85%.

This present report outlines our preliminary experience with combination of radiation, Mitomycin C and 5FU infusion in the treatment of colorectal adenocarcinoma and esophageal squamous cell carcinoma.

MATERIALS AND METHODS

Between January and October 1984, we have treated 33 patients with concomitant radiation, Mitomycin C and 5FU infusion. Nine patients had unresectable esophageal carcinoma with dysphagia and obstruction. Seven were T_2 lesions, two were T_3 lesions. Sixteen patients had recurrent colorectal adenocarcinoma with severe pain, bleeding or obstruction. Eight patients received the combination therapy on an adjuvant basis after surgical resection. The latter group all had locally advanced rectal or rectosigmoid adenocarcinoma with regional lymphadenopathy and infiltration of surrounding omentum, pelvic sidewall, posterior bladder or vaginal wall. Five had pT4, two had pT3a, and one had pT2 lesions.

There were 20 males and 13 females in the group with a median age of 62 years. Performance status (ECOG Scale) was two or three for all 25 patients with recurrent colorectal carcinoma and esophageal carcinoma. The adjuvant group all had performance status of zero at presentation. Two in the esophageal carcinoma group and ten in the recurrent adenocarcinoma group had distant metastatic disease at presentation.

Treatment Regime

Treatment was given in two split courses, with a rest period of four weeks in between. Both radiation and chemotherapy were started on the same day. Radiation was delivered by 10 or 15 MeV photons, AP-PA parallel pairs, 2000 to 2500 cGy in 10 fractions over two weeks, calculated at mid-plane. Treatment portals covered both the tumor mass and potential extension with adequate margins.

Continuous 5FU infusion was given from day one to five, at 20 mg per kg per day for 120 hours. Infusion usually started four hours before the first irradiation treatment. A single bolus injection of Mitomycin C was given on day one of the first course, $8mg/m^2$. Mitomycin C was deleted in the second course to minimize potential myelosuppression.

All patients in the esophageal carcinoma group had pre-treatment esophagram and endoscopy. In the colorectal recurrent group, 13 patients had pretreatment CT scan, two had exploratory laparotomies, and one had palpable vaginal disease. Hemoglobin, WBC count and platelet count were monitored weekly.

Treatment response was assessed both subjectively and objectively. Complete subjective response is defined as complete resolution of presenting symptom(s) for two months, and partial response is improvement of symptom(s) for two months. (For partial pain response, there must be at least 50% reduction in analgesic requirement). Objective response is assessed by radiological imaging or physical examination. Complete response means complete resolution of measureable or evaluable disease for two months, and partial response is 50% or more reduction of measureable or evaluable disease for two months.

RESULTS

Esophageal Carcinoma

Two patients had only one course of treatment. One developed uncontrolled hypercalcemia, and another developed a tracheo-esophageal fistula, likely from tumor necrosis. Both died with uncontrolled primary disease. Of the seven patients who completed the treatment, five had subjective CR and two had subjective PR. All had follow-up esophagram which showed opening of lumen with smooth stricture. Two patients had histological confirmation of fibrosis only in subsequent endoscopy. No recurrence of dysphagia has been observed at a median follow-up of six months.

Recurrent Colorectal Adenocarcinoma

Ten patients had completed two courses of treatment. Six patients had just one course because of previous irradiation in the same region. Pain palliation is good, with seven CR and eight PR. Only one patient showed no pain relief with treatment. Response usually occurred early, while patient was still on treatment. Seven patients had post-treatment reassessment. Six had CT scans: three had partial regression, two showed cystic changes within the mass without size change, and one showed progression. The seventh patient had a large vaginal mass. Only a plaque of induration was palpable at three months follow-up. The other nine patients did not have radiological reassessment because of rapidly progressing metastatic disease.

Adjuvant Therapy for Rectosigmoid Adenocarcinoma

So far no local recurrence has been observed. A median follow-up of five months is too short for analysis.

Toxicity

We found this combination well tolerated, even in the elderly. The oldest patient we have treated was 81 years of age. Treatment related side-effects include perineal skin reaction, diarrhea, stomatitis, esophagitis, nausea and hematological suppression. Six patients had mild erythema and two had patchy desqumation in the perineum. Two had mild stomatitis from 5FU infusion. Fifteen out of twenty-four patients who received pelvic radiation developed diarrhea, only five required parasympatholytic medication. Four had mild nausea. One patient with esophageal carcinoma developed tracheo-esophageal fistula after one course of treatment. None had developed significant myelosuppression that required interruption of treatment. Overall, only the patient with tracheoesophageal fistula had treatment interrupted because of toxicity.

DISCUSSION

Concomitant radiation with Mitomycin C and 5FU infusion is well tolerated with minimal acute toxicity. Habeshaw[8] and Danjoux[9] had reported increased late gastrointestinal toxicity, while in Wassif's series, only 10% had late toxicity. We need longer follow-up to determine our late toxicity rate.

This combination treatment provides good and prompt symptomatic palliation (pain, bleeding, dysphagia). In esophageal carcinoma, we have observed corresponding objective evidence of tumor regression. The picture is less clear with recurrent colorectal carcinoma. Symptomatic response has been excellent, but objective response criteria has to be established. Nine out of sixteen patients were not evaluated radiologically because of progressive distant metastatic disease. We plan to treat more patients with longer life expectancy in order to assess objective response, since adenocarcinoma may have slower tumor regression than squamous cell carcinoma. Since October 1984, we have treated another six esophageal carcinoma and thirteen recurrent colorectal adenocarcinoma patients. The acute toxicity and symptomatic palliation is very similar to our first group of patients.

The adjuvant role of concomitant radiation and infusional chemotherapy is unknown in locally advanced rectal and rectosigmoid carcinoma. We will continue our pilot study to test whether there is benefit with local disease control and possibly survival. We are encouraged by our initial studies which suggest this regime offers a palliation which is prompt, well tolerated and consumes little time for the patient.

We are indebted to our colleagues in the Gastro Intestinal Clinic, Tom Baker Cancer Centre for their assistance in this study.

REFERENCES

1. N.D. Nigro, V.K. Vaitkevicius, T. Buroker, G.T. Bradley, B. Considine, "Combined Therapy for Cancer of the Anal Canal", Dis. Colon Rectum, 24:73-75 (1981).

2. B. Cummings, T. Keane, G. Thomas, A. Harwood, W. Rider, "Results and Toxicity of the Treatment of Anal Canal Carcinoma by Radiation Therapy or Radiation Therapy and Chemotherapy", Cancer, 54:2062-2068 (1984).
3. J.E. Byfield, R.M. Barone, T.R. Sharp, S.S. Frankel, "Conservative Management Without Alkylating Agents of Squamous Cell Anal Cancer Using Cyclical 5FU Alone and X-ray Therapy", Cancer Treat. Rep., 67:709-712, (1983).
4. J.E. Byfield, R.M. Barone, J. Mendelsohn, S. Frankel, L. Quinol, T. Sharp, S. Seagren, "Infusional 5-Fluorouracil and X-ray Therapy for Non-resectable Esophageal Cancer", Cancer 45:703-708 (1980).
5. R. Franklin, Z. Steiger, G. Vaishampayan, I. Asfaw, J. Rosenberg, J. Loh, J. Hoschner, P. Miller, "Combined Modality Therapy for Esophageal Squamous Cell Carcinoma", Cancer 51:1062-1071 (1983).
6. T. Keane, A. Harwood, B. Cummings, G. Thomas, "Concomitant Radiation and Chemotherapy for Squamous Cell Carcinoma (SCC) Esophagus", Int. J. Radiat. Oncol. Biol. Phys. 10, Supp 2:89 (1984).
7. S.B. Wassif, "Ten Year's Experience with a Multimodality Treatment of Advanced Stages of Rectal Cancer", Cancer 52:2017-2024 (1983).
8. T. Habeshaw, J.S. Adams, "Weekly Large Fraction Radiotherapy and 5 Fluorouracil as a Palliative Treatment for Large Bowel Carcinoma: A Pilot Study", Int. J. Radiat. Oncol. Biol. Phys. 8:1127-1130 (1982).
9. C.E. Danjoux, G.E. Catton, "Delayed Complications in Colo-rectal Carcinoma Treated by Combination Radiotherapy and 5 Fluorouracil - Eastern Co-operative Oncology Group (ECOG) Pilot Study", Int. J. Radiat. Oncol. Biol. Phys. 5:311-315 (1979).

SECTION III: TECHNIQUES FOR THE ADMINISTRATION OF CHEMOTHERAPY AGENTS BY CONTINUOUS INFUSION

PROCEDURES FOR THE USE OF IMPLANTABLE AND EXTERNAL PUMPS FOR CONTINUOUS INFUSION CHEMOTHERAPY

Bettina Bentley Willis

Department of Nursing and Medicine, SUNY-Downstate Medical Center, Brooklyn, New York

The administration of chemotherapeutic agents is now being done more frequently on an outpatient basis which contributed to the improvement of the quality of life of cancer patients; this has become possible since small portable pumps were developed. The use of two of these pumps, one implantable (Infusaid), the other external (Cormed) will be described in this article. They are herein described, because of the author's familiarity with their use for more than three years at the Chemotherapy Unit of Downstate Medical Center, Brooklyn, N.Y. Several other types of portable pumps are also available and apparently equally effective, but the author did not have the opportunity to use them.

The Infusaid pump (Fig.1, see appendix) is an implantable apparatus which continuously infuses the desired dose of chemotherapy into the hepatic artery. It is implanted under the abdominal skin and therefore allows the patient complete mobility and minimal self-care. It does, however, require that patient returns to the clinic or physician's office for refills at regular intervals. The care, maintenance, and management of this delivery system are factors to be considered. To date, the use of Infusaid pump is approved by the FDA only for the administration of FUDR.

The following equipment should be assembled prior to the refilling procedure:

-Betadine swabs, alcohol swabs, surgical gloves, sterile drape, template (optional), huber needles, emptying syringe (without plunger), refilling syringe, refill tubing, 2" X 2" gauze, bandaid or spot.

The patient should be placed in a supine position. After identifying the pump site, its perimeter and side ports, it may be necessary to aspirate fluid, as the presence of a seroma surrounding the pump is frequently seen. Sterile techniques should be employed throughout the refilling procedure. The content of the seroma should be first removed in a syringe and should be appropriately cultured before proceeding with the routine of pump refilling. This routine should begin with the cleansing of the skin around and over the pump, first with betadine then with alcohol. A template may be used to locate the septum by placing it directly over the pump using the sideport as a guide. If flushing is needed this can be done through the sideport. In some cases, there may be two sideports, depending upon the type of pump which has been implanted (single or dual catheter).

The central rubber septum should be identified (it is located directly in the center of the pump). The refill tubing can then be attached to the emptying syringe, and a special needle is attached to the tubing. It is imperative that the integrity of the rubber septum be maintained; for this reason we recommend the use of the Infusaid huber needles to avoid any coring, thereby preventing leaks and subsequent complications (a refill kit containing all the necessary items can be purchased from the company). The pump should be held steady with one hand. The needle is inserted into the septum by injecting directly perpendicular into the pump (right angle). Upon opening the stopcock on the tubing, if the needle is in place, fluid should return immediately into the empty syringe.

Sometimes it may be necessary to "walk" the needle around under the skin, if the initial puncture is not on target. This prevents needle withdrawal, and saves the patient from the discomfort of another puncture "Walking" the needle clockwise subcutaneously frequently allows to successfully locate the septum without withdrawing the needle.

Once the needle is in place, and the fluid is returning, the syringe barrel is lowered to a level below the patient and maintained like this until the fluid has stopped rising in the syringe barrel. Once the meniscus ceases to rise in the barrel, the stopcock is turned off, preventing any of the fluid from returning into the pump inadvertently. The returned fluid must be measured, adding one ml to account for the fluid left in the tubing. The stopcock and syringe combination with the fluid can then be disconnected and replaced with the medication-filled syringe. The needle should not be removed. The medication can now be injected slowly with steady pressure. The plunger should be released every 5 cc, causing it to rise slowly, as the fluid returns to the syringe. This will confirm the fact that the needle is indeed still in place. One must never aspirate directly from the pump, as this may introduce blood into the pump and render it inoperable. Upon completion of the refill, the needle is removed with one swift upward thrust, and a 2 X 2 gauze is applied with slight pressure. When bleeding has ceased, a bandaid or spot is applied, and the refill procedure is completed.

It is of the utmost importance that patients keep their refill appointments, as the pump must never be allowed to empty completely. Patients should be instructed to return to the ambulatory clinic or physician's office for refill at 2-3 week intervals.

Each patient's pump has its individual table of curves supplied by Infusaid company to the surgeon at the time of implantation; they permit to calculate the FUDR dose and the administration flow rate.

While bearing the pump patients must avoid temperature extremes (e.g. sunbathing, saunas, etc.), high altitudes, and any activities which may cause trauma to the pump site. Such factors may cause a change in the pump flow rate; patients should notify their doctor or nurse if they become febrile, plan air travel or relocation to higher altitudes. In these cases the amount of drug and fluid injected in the pump should be recalculated applying special correction factors indicated in the instructions included with each pump. Should air travel be planned, the patient should be given an explanatory note, as the metal of the pump may cause questions at airport checkpoints.

The portable Cormed pump (Fig.2, see appendix) is an external device, which is attached to the patient's central venous catheter (e.g. the Hickman's and Broviac's catheters, etc.) and can infuse the desired dose of medication to the patient's circulatory system via a peristaltic pump action over a pre-set length of time.

Patient teaching should include care of the central venous catheter and care of the pump itself; the latter is minimal, however, patients should be taught to effectively troubleshoot the problems that may arise.

Each Cormed pump comes with the following accessories: drug reservoir bag with tubing; 2 rechargeable power packs; battery charger; specially designed screwdriver; flow rate meter and carrying case with shoulder/waist harness.

The pump and its accessories should be assembled, and the prescribed chemotherapeutic agent should be prepared. The sterile drug reservoir bag can be filled with the appropriate dose and volume for infusion (up to 60 cc/day). The tubing is securely attached to the reservoir bag, and primed by exerting slight pressure on the bag. The system should then be closed with a cap. Aseptic technique must be employed, and all tubing connections must be kept sterile.

The power pack is charged for 7 days use; it can be recharged over a period of 16 hours. Once the battery is connected to the pump, the flow rate can be determined by attaching the pump to the flow meter, turning the pump on, and adjusting the rate by appropriately turning the flow rate knob with a screwdriver. The pump can now be loaded with the medication to be administered.

The pump cover screws are removed, and the tubing is treaded into its track by utilizing the plastic tip of the screwdriver. A screw located inside the pump casing enables one to rotate the pump shaft for easy loading. The tubing must encircle the shaft in its track approximately twice to exit at the opposite notch from where it entered. The pump cover is replaced, the screws tightened, and the pump is closed. The tubing inside the pump must not be pinched. Now the pump can be attached to the patient's central venous catheter.

Air must not enter the patient's catheter; for this purpose the tube should be temporarily closed with a clamp or better, by bending it and pressing it tight between the operator's fingers; this latter manouver can avoid occasional tearing of the catheter caused by clamps. The tip of the catheter is cleaned with an alcohol swab, and its cap is removed. The luer lock connection of the Cormed tubing is then attached to the catheter, the pump is turned on, and the procedure is completed.

Patient teaching should include:

(1) How to turn the pump off and on.
(2) How to notify the nurse or doctor should the constant humming of the pump stop.
(3) How to proceed in case of a backflow of blood is noted; this may indicate some malfunctioning. The tubing may be pinched inside the pump casing; it can be checked by the patient. If the catheter is clotted, the nurse or doctor should be notified. They may attempt to lyse the clot using urokinase or streptokinase;
(4) The handling of leaks or air bubbles; the patient should make sure that all connections are tight; if more than one air bubble is present in the tubing or leakage continues, the patient should turn off the pump and disconnect it from the catheter, and should report to his physician or nurse.
(5) How to protect the pump from getting wet: the pump should be removed prior to bathing, wrapped in plastic and set outside the bath tub on a chair or table.
(6) To notify his nurse or physician of any fever over 100.5 F. lasting more than 24 hours.

Patients should also be informed of the potential side effects of specific chemotherapeutic agents delivered through the pumps, the measures he/she may take to counteract them, and the need to notify their physician immediately in case more severe ones develop (i.e.: bleeding, chills, etc.).

We hope that these guidelines could assist in the goal of optimizing the care for patients receiving continuous infusion chemotherapy.

REFERENCES

1. Cozzi, E., et al. Nursing management of patients receiving hepatic arterial chemotherapy through an implanted infusion pump. Cancer Nursing, 229-234, June 1984.
2. Goodman, M.S. and Wickham, R. Venous Access Devices: an Overview. Oncology Nursing Forum, 11:5, 16-23, Sept./Oct. 1984.
3. Winters, V. Implantable vascular access devices. Oncology Nursing Forum, 11:6, 25-30, Nov./Dec. 1984.

APPENDIX

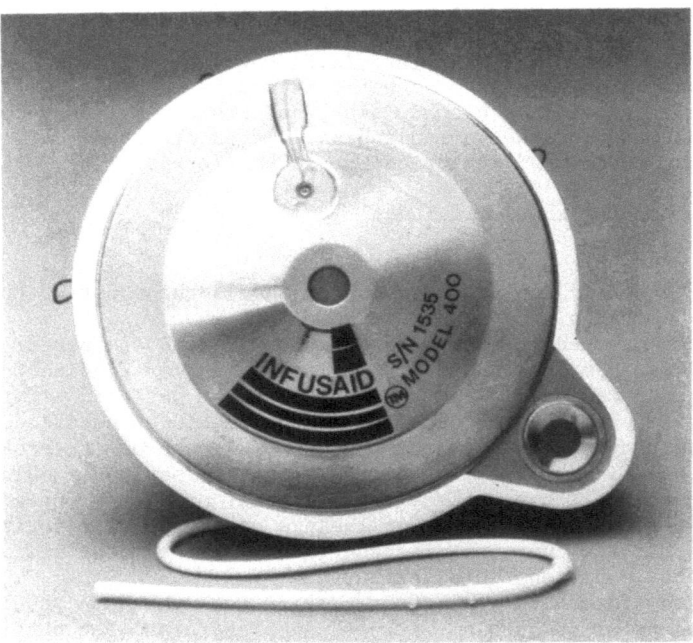

Fig. 1. The implantable Infusaid pump (Model 400); exterior of its upper part.

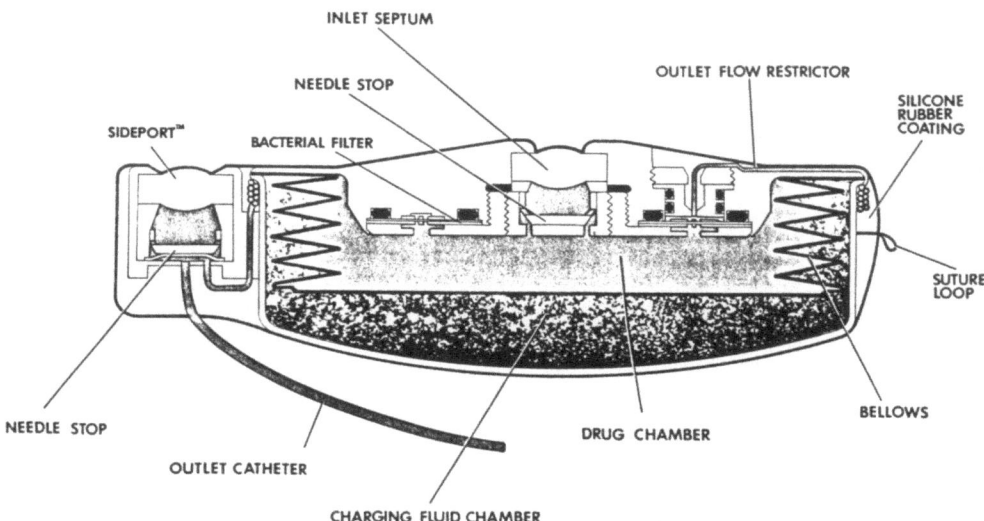

Fig. 2. The Infusaid pump (Model 400 - sagitol section: it is a titanium cylinder in which the changing fluid is sealed between bellows and the shell. When filled with medication and/or heparin and sterile water, the charging fluid is compressed into a liquid. As the patient's body temperature warms the charging fluid, it becomes a vapor. This vapor pressurizes the refill fluid thereby releasing it from the drug reservoir.

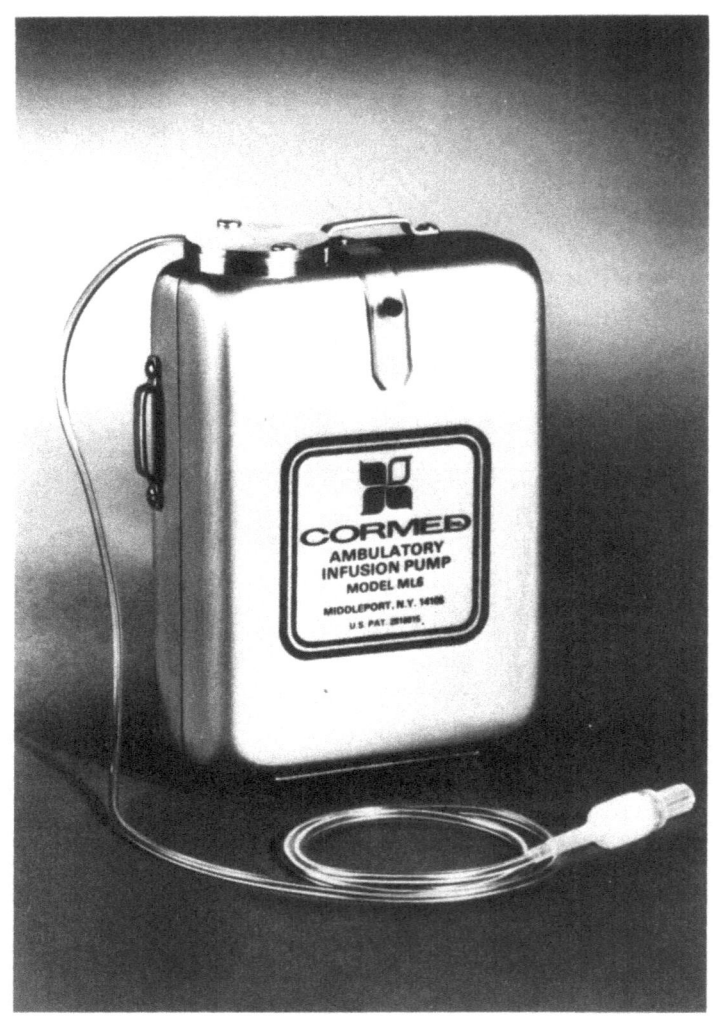

Fig. 3. The portable Cormed pump, outside presentation

CENTRAL LINE CATHETER CARE: THE NURSE AND

THE PATIENT'S PERSPECTIVE

        Mary Jane Tunny

        Downstate Medical Center
        450 Clarkson Avenue
        Brooklyn, N.Y. 11203

Vascular access in patients requiring antineoplastic drugs is often a serious and challenging problem. Chemotherapy protocols frequently involve drugs that are severely irritating to peripheral veins. As a result patients receiving I.V. antineoplastic drugs over months or years experience thrombosis, sclerosis and destruction of available surface veins. The search for a suitable vein can become a painful ordeal for the patient, a time-consuming and disruptive endeavor for the physician and oncology nurse.

Central venous catheters (Hickman) have received widespread acceptance over the past decade and have been a major breakthrough in the management of patients requiring long term therapy. The Hickman Catheters are unique among central venous catheters because they are made of silicone rubber and are available in single, double, and triple lumen. The catheter has a small dacron cuff that causes a fibroblastic reaction following insertion, which stabilizes the catheter and is thought to provide some barrier against bacterial invasion from the exit site.

The Hickman Catheter provides a venous access for the administration of I.V. antineoplastic drugs, blood products, antibiotics, parenteral nutrition and venous sampling both in the hospital and home setting.

Nursing management of the patient with a Hickman Catheter is threefold and consists of procedures for catheter care, education of patient and family and monitoring for complications.

Patient education should include a written instruction booklet that lists step by step directions for each aspect of catheter care. The booklet should include such topics as general information about the catheter, catheter exit site care, dressing change, heparinization and catheter capping. The principles of aseptic technique and air embolus should be incorporated, as well as providing a teaching tool. This booklet will serve as a reference manual for the patient at home.

The educational process begins prior to the insertion of the catheter and continues post insertion. This includes written information as well as demonstration by the oncology nurse of each aspect of care. Visual aides are used for instruction. Patients are required to demonstrate their ability to manage all procedures prior to discharge.

In order to prevent infection at the exit site, patients are instructed to change the dressing three times a week using clean technique. Observation of the exit site for increased redness, tenderness and drainage are noted. The catheter is cleansed with alcohol if any cellular debris is present. The manufacturer of the catheter strongly discourages the use of acetone and iodine tincture in the care of the catheter. The exit site is cleansed three times with povidone-iodine working outward in a concentric circle. Following the cleansing procedure providone-iodine ointment is placed at the catheter exit site and an occlusive dressing (Tagaderm or Opsite) applied. This type of dressing permits observation of the exit site at all times and provides comfort for the patient. The remaining portion of the catheter is taped with micropore tape to the chest wall to prevent tension on the catheter.

To avoid air embolus and hemorrhaging, all Hickman Catheters are capped with an intermittent luer lock infusion cap. This type of cap allows the patient to irrigate the catheter without removal of the cap and is changed weekly using sterile technique. When the cap is changed the patient is advised to clamp the catheter with a smooth edged or rubber-shod clamp. Following clamping of the catheter the cap is removed, catheter cleansed with providone iodine and the sterile infusion cap applied. Patients are required to carry clamps at all times as an emergency precaution.

When catheters are not in use for continuous therapy, they are irrigated three times a week with heparin to preserve their patency. The amount of heparin used varies from 10u/ml to 100u/ml depending on the physician order. The volume of solution required to keep the catheter patent is 1.8 ml to 2.5 ml. A short 25 gauge needle is attached to the heparin syringe, the cap is cleansed with providone iodine and the prescribed amount of heparin injected. Catheter irrigations are required following any use of the catheter and blood sampling.

Patients are advised to contact their physician if they are unable to flush the catheter or meet with resistance during the procedure. Injection of a dilute solution of a potent fibroinolytic agents, (urokinase or streptokinase), into the central venous catheter consistently reestablishes its patency.

Occasionally a leak or breakage of the catheter may occur if the catheter is punctured with a needle. If this should occur, patients are advised to clamp the catheter with the smooth edged occlusion clamp at the point between the exit site and the damaged catheter and notify physican. Catheters can easily be repaired with the repair kit provided by the manufacturer and used twenty-four hours later.

Hickman catheters are still an unknown entity to the general public and even some medical personnel. For this reason patients should be encouraged to wear a Medic Alert bracelet or carry an information card describing the catheter and the location.

Referral to the local pharmacy is required for equipment needed in the management and care of the catheter. Referrals to the Visiting Nurse Association have proven invaluable in assisting the patient and the family in the adoption of hospital learned catheter care in the home setting. The patient should be given a list of resource medical personnel who can be contacted at any time should questions arise or in case of an emergency.

Follow up visit should be scheduled in the hospital or clinic by the oncology nurse to evaluate the patients' progress and adherence to the principle and practices of catheter care which is vital to the prevention of complications associated with Hickman catheters.

The benefits of the Hickman Catheters include, ready access, ease of chemotherapy without extravasation, decreased time spent in administration, and less trauma for the patient and staff. The long term use of Hickman Catheters is safe, practical and provides considerable comfort for the oncology patient requiring long term therapy in the hospital or home.

REFERENCES

Anderson, M., Aker S., Hickman R.: The Double Lumen Hickman Catheter, American Journal Nursing 82 (2), 272-274, 1982.
Carelli, R., Hernik, E.: Hickman/Broviac Catheters, Result of Survey and Patient Care Considerations, NITA Vol. 7. July/Aug 9, 1984.
Goodman, M., Wickman, R., Venous Access Devices: An overview, Oncology Nursing Forum vol. 9, No. 4, 1982.
Schneider, P., Unclotting Subclavian Vein Catheters, Infusion, Vol. 7, No. 12, Nov/Dec. 1982.
Schaefer, N., Hickman Catheters, NITA Vol. 4, July/Aug. 1981.

POTENTIAL COMPLICATIONS OF

RIGHT ATRIAL CATHETERIZATION

>Richard M. Stillman
>
>Department of Surgery
>SUNY - Downstate Medical Center
>Brooklyn, NY

INTRODUCTION

Often credited with the introduction of the use of central venous catheters in humans, 1956 Nobel laureate Werner Forssman stressed acquiring a functional knowledge of anatomy, application of gentle technique, and selection of the proper catheter in order to avoid numerous obvious and potential complications. But even today, almost 30 years later, human error remains the single most frequent cause of complication in the widespread use of central venous catheters.

CLASSIFICATION & INCIDENCE

The overall complication rate in the insertion and use of central venous catheters is about 14%. This includes an approximately equal distribution of major mechanical complications, minor mechanical complications and infectious complications. Management of these complications adds a cost to hospitalization exceeding that of the uncomplicated insertion of the other 86%! This is another illustration of Francis D. Moore's perceived "high cost of low frequency events".

MECHANICAL COMPLICATIONS

<u>Malposition</u>. The long duration of use of the right atrial catheter increases the chances that a malpositioned catheter tip will result in major morbidity. Misplacement by overadvancent may lead to delayed vascular perforation by pressure necrosis of superior vena cava or innominate vein, venous thrombosis, cardiac arrhythmias or perforation with tamponade, hydromediastinum, hydrothorax, or coronary sinus thrombosis. Altered blood flow patterns with knots, loops or late displacement into such shapes may result in swelling, pain and venous prominence of the arm, shoulder, face or supraclavicular area. Presumably the Dacron cuff tends to retard late catheter tip migration (up to 10 cm with subclavian vein and 3 cm with internal jugular) as is seen with other central venous lines. Guesses about posiiton based on the external body surface are notoriously misleading. Therefore, fluoroscopic confirmation of position at the time of insertion should be sufficient to confirm

adequate position. This ideal position is in the upper segment of the superior vena cava above the sagittal Xray plane drawn thru the third rib, T4-5 interspace, or carina; i.e. three to four cm above the superior vena caval-right atrial junction.

<u>Catheter breakage; embolization</u>. Typically due to design defects with the older central venous catheters, Hickman catheter breakage usually occurs either at the catheter to hub junction, or at a point where a traumatic clamp was used to occlude flow. Even the supposedly safe atraumatic clamp marketed for use with Hickman catheters will cause catheter performation if used frequently. Use of the rubber Gelco cap allows needle pucture for infusion, blood aspiration and heparin irrigation without the use of catheter clamping. If despite this measure, breakage occurs, a repair kit is marketed which allows resection and replacement of the proximal end of the catheter by use of cement and a sterile sleeve. Breakage during insertion is avoided by use of a venous cutdown or application the Seldinger technique for insertion. Thru-the-needle catheters should never be used. A patient's sudden unexpected movement may cause breakage and embolization.

<u>Catheter thrombosis</u>. Thrombosis occurs because of blood reflux into the catheter tip. This may result from cessation of infusion, kinks in the line or transient increases in central venous pressure. Thrombosis is best avoided by flushing the catheter after each use with heparinized saline (10 units per ml; about 20 ml). Prolonged infusions should contain 0.5 to 1.0 units of sodium heparin per ml of solution. This has been shown to decrease catheter thrombosis and catheter-related sepsis. Minute aliquots of streptokinase or urokinase infused over several hours may reopen a thrombosed catheter.

VASCULAR

<u>Endothelial damage; venous thrombosis</u>. Prolonged use of indwelling venous catheters not unexpectedly produce erosion of vascular endothelium. Venous thrombosis may be a long-term result.

<u>Chemical phlebothrombosis</u>. Presumably, appropriate positioning in high flow central veins lessens the incidence of chemical phlebothrombosis due to infusion of irritating materials.

<u>Superior vena caval perforation</u>. The high morbidity from this complication suggests use of a smooth catheter tip, not cut at an acute angle despite the temptation to provide a tapered end to facilitate insertion.

<u>Arterial damage</u>. Arterial spasm, hand ischemia, carotid artery cannulation are potential complications of the Seldinger technique of insertion. One must be certain that the initial needle stick enters a vein, otherwise the subsequent wire and large bore catheter may well enter an artery of comparable diameter. Arterial puncture should be relatively innocuous if the initial needle is removed and gentle pressure applied until well after hemostasis is achieved. This may be somewhat of a problem in thrombocytopenic patients.

CARDIAC

<u>Air embolism</u>. "Air fluxes into veins as surely as it enters the alveoli." Entry rates of 1 cc/kg/minute are fatal in animals. In man,

Table. The most common complications of central venous catheter insertion.

COMPLICATION	FREQUENCY	MANAGEMENT	PREVENTION
Catheter-related Sepsis	7%	Fever workup; Catheter removal; Antibiotic	Sterile technique in insertion and in use of line; Small amount of heparin in line
Malposition	2.3%	Replace or reposition catheter	Use fluoroscopy during catheter insertion
Pneumothorax Hemothorax	1.4%	Tube thoracostomy	Internal jugular is safer than subclavian; Know anatomy
Vascular perforation	1%	May require surgical repair	Fluoroscopic placement of catheter tip in upper superior vena cava; Smooth catheter tip
Thrombosis	0.7%	Heparin or kinase infusion; Replace catheter	Avoid cessation of infusion; Heparinize catheter
Air embolism	0.5%	Cap hole; Trendelenberg & left side down; Aspirate air; Cardiopulmonary resuscitation	Trendelenberg position while disconnecting and inserting catheter
Catheter breakage	0.3%	Repair or replace	Avoid traumatic clamping of catheter; Use rubber cap; Better designed catheter especially at shaft-hub junction
Catheter embolism	0.2%	Retrieval by interventional radiology or surgically	Use only over-the-wire (Seldinger) or cutdown technique for introduction of catheter.

20 cc/second causes symptoms, 75 to 100 cc/second is fatal. One hundred ml per second easily goes through a 14 gauge needle with a pressure gradient of only 5 cm water. Air entry via a hole in the vein wall or an open infusion system is facilitated by cyclical negative pressures generated within the thorax of the awake, spontaneously breathing patient undergoind Hickman catheter insertion under local anesthesia, or subsequent use of the catheter. This is especially hazardous if the patient is sitting, tachypneic or hypovolemic. A mechanical ventilator may have similar aggravating effects. Oter potential problems are design defects in the line with stopcock fracture or wrong position, line fractures, hub fractures. A confused patient may tamper with the system. Therefore, as simple and secure a system as possible is used when infusing via central lines. During insertion, prevention includes use of 20 to 30 degrees of Trendelenberg during insertion, and constant awareness of potential 'cardio-atmospheric fistulas'. The scrub nurse is instructed never to deliver an uncapped catheter to the surgeon. During critical points of insertion, such as when the Hickman is introduced through the sheath, the patient is instructed to Valsalva. Central venous pressure should not be excessively low during insertion. Any sudden alteration in consciousness, respiratory or CNS function suggests air embolism. A 'mill wheel' murmur may be audible, but time should not be taken to elicit confirmatory signs. Immediate placement in left lateral decubitus with head down will allow displacement of air from the pulmonary outflow tract. Then, an attempt should be made to aspirate air from the CVP line as it is advanced as far towards the outflow tract as possible. Cardiopulmonary resuscitation, hyperbaric oxygen, thoracotomy are considerations in severe cases. The mortality rate in air emolism is 29%.

Other cardiac complications. Coronary air embolism, right heart perforation, pericardial tamponade, acute right heart failure, arrhythmia, right atrial thrombus, right-sided endocarditis have been reported. These result from improper placement of the catheter, improper design or obliquely transecting the catheter tip, or infection.

PULMONARY

Pneumothorax, hemothorax, hydrothorax, pulmonary artery rupture, hemoptysis, tracheal puncture, retropleural hematoma are rare complications usually due to problems with catheter insertion.

NEUROLOGIC

Cerebrovascular accident, paradoxical air embolism to brain, Horner's syndrome, accessory nerve paralysis, brachial plexus injury, phrenic nerve palsy, reversible hydrocephalus have each been reported as coomplications of central venous access. Careful palpation of adequate veins, throough knowledge of regional anatomy, Doppler auscultation especially in locating the subclavian vein, and fluoroscopy to confirm proper position are suggested measures to prevent these problems.

INFECTIOUS

Bacteremia, catheter-related sepsis. Central line insertion places the skin surface and vascular endothelium is continuity, effectively bypassing normal defenses against bacteremia -- skin surface, lymphatics, vessel wall. The incidence of catheter-related sepsis is about 7%, but increases with time. Overall reported rates of catheter-related sepsis range from 4.1% to 39.8%. Prevention is by meticulous aseptic technique in insertion, use and maintenance of catheters. "Stopcocks do not stop cocci from entering the circulation and side ports provide a fine harbour for microorganisms." A small physiologic concentration of heparin added to any intravenous solutions which are to be infused over a period of hours will decrease clot formation at the catheter tip. Such clot is a necessary (though not sufficient) condition for catheter infection, providing a nidus for bacterial growth.

Suppurative phlebitis. Infected veins may need to be resected.

Clavicular osteomyelitis. This is an exceedingly rare complication of subclavian catheterization.

DYE-RELATED

Acute tubular necrosis, anaphylaxis. One must not forget the small volume of dye which is sometimes needed during catheter insertion to opacify the catheter or delineate venous anatomy. The use of radioopaque catheters is encouraged.

SUMMARY

Overall, one must remember Werner Forssman's guidelines for avoidance of central venous catheter complications.

[1] Acquire a sound knowledge of anatomy

[2] Select the right catheter (and attached equipment)

[3] Use appropriately gentle and sterile technique

Currently the risk of potentially fatal events in right atrial catheterization is 1:100. This is one-hundred times larger than the 1:10,000 annual per capita risk of death in an automobile accident. There is obviously room for improvement.

REFERENCES

Coppa GF, Gouge TH, Hofstetter SR: Air embolism: a lethal but preventable complication of subclavian vein catheterization. JPEN 1981: 5: 166-168.

Eisenhauer ED, Derveloy RJ, Hastings PR: Propective evaluation of central venous pressure (CVP) catheters in a large city-county hospital. Annals of Surgery 1982; 196(5): 560-564.

Peters JL: Current problems in central venous catheter systems. Intensive Care Medicine 1982; 8: 205-208.

Vanholder R, Lameire N, Verbanck J, van Rattinghe R, Kunnen M, Ringoir S: Complications of subclavian hemodialysis: A 5-year prospective study in 257 consecutive patients. International Journal of Artificial Organs 1982; 5(5): 297-303.

LONG TERM COMPLICATIONS OF THE

INDWELLING CENTRAL LINE CATHETERS

>Jose R. Marti
>
>Assistant Professor, Dept. of Surgery
>State University of New York
>Downstate Medical Center
>Brooklyn, N.Y. 11203

INTRODUCTION

The ability of providing a route for continuous infusion and the delivery of medications on an outpatient basis has been made possible by the catheters designed by Broviac and Hickman, initially utilized for patients requiring long term parenteral nutrition.(1)

This technique of subcutaneous central vein puncture and catheterization has been recently adopted for continuous infusion of hypertonic solutions, blood products and medications, and is becoming particularly popular for the administration of chemotherapy. This procedure decreases the incidence of complications caused by extravasation of cytotoxic substances in the soft tissues, and also the progressive scarcity of peripheral veins caused by the highly sclerosing nature of these solutions. It also allows for a continuous infusion of chemotherapy; and for prolonged treatments on an outpatient basis. However in return other set of complications can be expected among these patients. These complications can be minimized by careful attention and a meticulous technique in the maintenance of the catheter. The factors leading to complications can be classified into four broad categories:

1. Complications related to mechanical aspects.

    (a) clotting of the catheter
    (b) severance of the line
    (c) pump failure

2. Complications related to sepsis.

    (a) exit site
    (b) septicemia

3. Thromboembolic complications.

    (a) air emboli
    (b) mural thrombi

4. Complications related to product administration.

   (a) wrong dose
   (b) wrong product
   (c) wrong combination

Most of these complications can be avoided, the majority of them can be corrected without removing the catheter. Occasionally, however the catheter will have to be removed, and a new one reinserted. A catheter related death occurs rarely. In general, the minimum stay of the catheters is 3 months, and the average stay is reported to range between 9 - 30 months (1 & 2).

1. Complications related to mechanical aspects:

   (a) Clotting of the catheter.

This is perhaps one of the most common complications but it will rarely lead to a serious consequence. It usually is due to clot formation in the tip of the catheter and at least in one series it constituted the most common reason for removal of the line (2). Attention to a proper placing of the heparin lock will markedly reduce the incidence of catheter clotting.

The technique to flush the catheter that has already clotted is described elsewhere in this publication.

   (b) Severance of the line.

This is often caused by the patient repeatedly clamping the line over the same place. There are new catheters that have a double lumen which prevent backflow and reportedly even air emboli, however they are new and more experience is required (3). Another chapter in this subject describes precisely the proper way to handle this correctly and minimize the incidence of line breakage.

If the line breaks, most companies provide a repair kit which explain the repair procedure carefully.

   (c) Pump failure.

This complication is usually related to the actual mechanical functioning of the pump and though it should be taken into consideration a comprehensive explanation escapes the scope of this chapter.

2. Septic Complications:

   (a) Septicemia.

Often patients receiving chemotherapy are already immunosuppressed and are highly susceptible to infections if they do not already have one. The diagnosis of catheter related septicemia is not an easy one. It implies that blood cultures obtained through the catheter and through a separate distant venipuncture grow the same organisms which in turn will be the same that grows on the tip of the catheter. If the diagnosis is strongly suspected the only safe way to approach this, is by removing the catheter and placing a new one, if possible 24 hrs. later and ideally on the contralateral side (2).

(b) Infection at the exit site.

This is not an uncommon occurrence. It is usually caused by Staph epidermidis, and even though it is not as dramatic in its clinical presentation, it will often lead to catheter removal, despite proper antibiotic trial (2).

3. Thromboembolic Complications:

(a) Air emboli.

Even though this complication is frequently mentioned, it has been probably overrated in its incidence if not in its consequences as well. Proper connections to the solution bottles or to the heparin locks are essential, and negligence to achieve this can prove fatal.

(b) Mural emboli.

This is undoubtedly a more frequent complication than the latter one; and unfortunately one that requires removal of the catheter. It usually manifests itself as limb swelling, and a radiological study with contrast will diagnose it invariably.

The combination of sepsis and central vein thrombosis is almost always fatal.

Despite extensive research for less thrombogenic catheter materials and proper management, unfortunately this complication is rather unpredictable and difficult to prevent.

4. Toxic Complications:

This group of complications is rare, and it is almost always preventable since it is often related to human error.

A list of groups of medications or products that are commonly delivered through central catheters follows:

Antineoplastics	Miscellaneous:
Antibiotics/antifungal agents	. antinauseants
Blood products	. analgesics
Nutritional products	. sedatives
	. anticonvulsants

It is thus easy to imagine how the wrong dose, wrong product or the wrong combination can be administered, and one wonders why it does not happen more often.

Good rules of thumb include to double check on the orders for the product and dosage, and to avoid giving drugs in combination simultaneously.

Long term indwelling cathetorization provided the following advantages:

(a) Avoids sclerosing of periferal veins.

(b) Prevents extravasation of product into soft tissues.

(c) Allows for continuous infusion of chemotherapy delivery.

(d) Permits for complete outpatient treatment.

With the proper precautions, supervision and training the benefits far outweigh the risks.

1. Heimback, D.M., and Ivey, T.D. Technique for placement of a permanent home hyperalimentation catheter. Surg.Gynecol.Obstet. 143:634-636, Oct. 1976.
2. Riella, M.C., and Scribner, B.H. Five years' experience with a right atrial catheter for prolonged parenteral nutrition at home. Surg.Gynecol.Obstet. 143:205-208, Aug. 1976.
3. Bjeletich, J. and Hickman, R.O. The Hickman Indwelling Catheter Amer.J.Nurs. 80:62-65, Jan. 1980.

INDEX

Actinomycin D, 55, 183
Adenocarcinoma
  colorectal, recurrent, 218
  rectosygmoid, 219
Administration, intracavitary, 3
Adriamycin, 19-25, 54, 82, 83,
      87-90, 131, 158
  and biopsy, endomyocardial, 19
  cardiotoxicity and schedule, 23
  and childhood hepatic malignancy,
      87-90
    efficacy, 87-90
    toxicity, 87-90
  -cisplatinum combination, 88
  and congestive heart failure, 19-21, 24
  -cyclophosphamide combination, 22
  and endomyocardial biopsy, 19
  and heart failure, congestive, 19-21, 24
  and hepatic malignancy of childhood,
      see childhood hepatic
      malignancy
  and hepatoblastoma, 87
  and hepatocellular carcinoma
    infusion continuous of, 19-25,
      87-90, 183-188
  and malignancy, hepatic, see childhood hepatic malignancy
  and neutropenia, 89
  and radiation, concomitant, 183-188
  and soft tissue sarcoma,
      pleomorphic, 183-188
  -vincristine combination
    and esthesioneuroblastoma, 92
    and hepatoblastoma, 91, 92
  infusion, continuous, 91-94
  and medulloblastoma, 92
  and neuroblastoma, 92
  and neuroectodermal tumor, 91, 92
  toxicity, 92-93
    hematological, 93
  and Wilm's tumor, 91, 92
  weekly schedule best, 21-22
Amylase, 55
Anal cancer, squamous, 120
  and 5-fluorouracil, 120
  anal canal carcinoma, 135
  and FUMIR, 135-139, see FUMIR
  management, 136

Anal cancer, squamous
  anal canal carcinoma (continued)
    and metastasis, occult, distant,
      138
    survival rate, 135
    treatment schedule, 135, 137
ANLL, see Leukemia, acute non-lymphocytic
Anthracycline, see Doxorubicin
Antidote, 3
Antigen, carcinoembryonic (CEA), 74
Ara-C, see Cytarabine
Artery, hepatic infusion therapy
      (HAI), 195-199

Bile duct stricture, 73
Bladder cancer, invasive, 155-157
  cell carcinoma, transitional, 155
  conduit, ileal, 155
  cystectomy, radical, 155
  incidence in the U.S.A., 155
  resection, transurethral, 155
  transitional cell carcinoma, 155
  transurethral resection, 155
  urologist's
    complications, 156
    problems, 156
Bladder carcinoma, 149-153
  chemotherapy by infusion, 149-153
  combined therapy, 149-153
  cure rate, 149-151
  cystectomy, 149
  death rate in the U.S.A., 149
  infusion chemotherapy, 149-153
  radiation, 149-153
  rate of
    cure, 149-151
    death, 149
    survival, 149
  survival rate, 149
  toxicity, 150
Blenoxane, see Bleomycin
Bleomycin, 5-6, 13-18, 81
  blood level, 14
  and cervical cancer, 16
  and DNA, 5-6
  efficacy, 6
  and fibrosis, pulmonary, 5, 6, 15

Bleomycin, (continued)
  half life, 14
  infusion, continuous, 5-6, 13-18, 81-83
  and pulmonary fibrosis, 5, 6, 15
  therapeutic index, 5-6
  toxicity, 6, 13, 14, 83
  -vinblastine combination, 6
Breast cancer patient, 54, 56

Cancer
  anal
    epidermoid, 133
    squamous, 120
  bladder, invasive, 155-157
  breast, 54, 56
  cervical, 16, 120
  colorectal, 27-42, 61, 127, 217-220
    combined therapy, 217-220
    and 5-fluorouracil infusion,
        continuous, 217-220
      response rate only 20%, 28
    in liver, 120
      and 5-fluorouracil, 120
    a major problem, 37
    and mitomycin C infusion,
        continuous, 217-220
    patient, 55-58
    primary, 127
    radiation, concomitant, 217-220
    toxicity, 219
    transplantable in mouse, 32
    treatment regimen, 218
  esophageal, 118-120, 207-209, 217-220
    and cisplatinum, 119
    and combined therapy, 207-209, 217-209
      5-fluorouracil, 118-120, 207-209
      mitomycin C, 207-209
      radiation, concomitant, 119, 207-209
    and 5-fluorouracil, 118-120, 207-209
    and mitomycin C, 119, 207-209
    and radiation, 207-209, 217-220,
    survival, poor, 119, 207
    toxicity, 219
    treatment regimen, 218

  gastrointestinal and radiation,
      217-220
  head and neck, 116-118, 189-193
    and 5-fluorouracil, 116-118
    and mitomycin C, 118
    of paranasal sinus, case report,
        189-193
      case report, 190
    and radiation, 117
    survival, 118
    toxicity, 117
  hepatic, 67, 87-90
    and adriamycin, 54, 87-90
      effective therapy, 89
      infusion, continuous, 87-90

Cancer (continued)
  hepatic (continued)
    and catheter in hepatic artery,
        54
    chemotherapy, preoperative, 89
    of childhood, 87-90
    dearterialization, 59-60
    and 5-fluorouracil, 54
    metastatic, 51-66
    and microsphere therapy, 67-69
    pre-operative test, 53
    starch microsphere, 51-56
  prostatic, 155
  and radiation therapy, 133-147
  rectal, 195
  squamous cell -, 133-147
  and treatment, combined, 201-205
Carcinoma
  of anal canal, 135-139
  of bladder invasive, see Cancer
      of bladder
  brochogenic, small-cell -, 211-216
    and cisplatinum, 211-216
    combined therapy, 211-216
    patient, 213
    and radiation, 211-216
    survival time, 213
    VP16, 211
  cervical, 142-143
    and FUMIR, 142, see FUMIR
    and pelvic failure, 142
    toxicity, 143
    treatment schedule, 143
  colorectal
    and FUDR, 73-78
    infusion
      intrahepatic, 73-78
      systemic, 73-78
    metastatic, 73-78
    hepatic, 195
      therapy, 127-132, 195
  esophageal, squamous cell, 217-220
  head and neck, 141-142
    and FUMIR, 141, see FUMIR
    and mucositis, 141
    recurrence-free, rate of, 142
  hepatocellular, 87
    and adriamycin, 87-90
  of paranasal sinus, 189-193
    case report, 190
    cisplatinum infusion, 189-193
    combined therapy, 189-193
    radiation, concomitant, 189-193
  rectal, 120-121
    and 5-fluorouracil, 120-121
    and mitamycin C, 121
  small-cell, bronchogenic, see
      bronchogenic
  transitional cell (TCC), 155
    in bladder, 151

Carbamazepine, 93
Cardiomyopathy, 19
Cardiotoxicity, 23
  and anthracycline, 19
  and biopsy, endomyocardial, 19
Carmustine, 59, 68
Catheter
  in artery, hepatic, 54
  central venous, 229-231
    nursing management of, 229-231
    patient training, 229-231
  complication, 239-242
    mechanical, listed, 239, 240
    septic, listed, 239-241
    thromboembolic, listed, 239-241
    toxic, 241
  good rules, 241-242
  indwelling, 239-242
Catheterization, right atrial, complications of, 233-237
  breakage of catheter, 234
  damage
    cardiac, 236
    neurologic, 236
    pulmonary, 236
  error
    human, 233
    mechanical, 233
  list of, 235
  sepsis, 237
  thrombosis, 234
  vascular damage, 234
CEA, *see* Antigen, carcinoembryonic
Cervix, cancer of, 16, 120
  and 5-fluorouracil, 120
  and mitomycin C, 120
  and radiation, 120
Chemotherapy and radiation, combined, 211-216
Chemotherapy, regional, 67
  advantage, 67
  combined, intra-arterial, 51-56
  continuous infusion, 3-12
  intra-arterial, 67
  in liver cancer, 67
Cis-diamminedichloroplatinum, *see* Cisplatinum
Cisplatin, *see* Cisplatinum
Cisplatinum, 6, 9, 119, 122, 141, 149
  bolus injection undesirable, 104
  combined therapy, 177-181, 189-193
  and DNA, 177
  infusion, continuous, 43-50, 101-105, 177-181, 189-193, 211-216
    and radiation, concomitant, 211-216
    in paranasal sinus carcinoma, 189-193
    remission rate, 179

Cisplatinum
  infusion, continuous (continued)
    side effect, 179
    in small-cell bronchogenic carcinoma, 211-216
    sirvival rate is low, 179
C-MOPP (cyclophosphamide, oncovin, procarbazine, prednisone), 79
Colorectal cancer, 27-42, 61, 127, 217-220, *see* Cancer, colorectal
Combination therapy, 3, 27-42, 133-153, 177-193, 201-205
Conduit, ileal, 155
Congestive heart failure, 19-21, 24
COPBLAM (cyclophosphamide, oncovin, prednisone, bleomycin, adriamycin, matulane (procarbazine)), 79
  infusion chemotherapy, 79-86
  and large-cell lymphoma, 79-86
  regimen, 82, 83
    cure rate, 83
    expensive, 83
    schedule, 80
    side effects, listed, 86
CORMED (external pump), 223-228
  ambulatory infusion pump, 223-228
  patient teaching, 225-226
  photograph, 228
  procedure for attachment, 224-225
CYVADIC regimen, 22
Cyclophosphamide, 8, 22, 82, 83
Cystectomy, 149, 150
  radical, 155
    mortality rate, 155
Cytarabine, *see* Cytosine arabinoside
  and DNA, 6
  and DNA polymerase, 6
  drug combination, 7
  and leukemia, acute, 6-7
  toxicity, 7
Cytosine arabinoside (Ara-C), 95-100
  infusion, continuous, 95-100
  and leukemia, acute, non-lymphocytic, 95-100
  and marrow response, 97
  and myelodysplastic syndrome, 95-100
  pharmacokinetics, 98-99
  and preleukemia, 99
  toxicity, serious, 98
Cytoxan, 22

Dearterialization, 59-60
DEF, *see* Dose effect factor
DNA, 5-6, 8, 27, 129, 177
Dose effect factor (DEF), 158, 161, 167
  definition, 158

Doxorubicin, 7-9, 158-176, *see also* Adriamycin
  and advanced malignancy, 158-176
    antitumor activity, 165
    drug response, 161-162, 164
    patient population, 160, 171
    pharmacokinetics, 162, 163
    response rate, 175-176
    side effects, 162, 164, 174
    therapeutic regimen, 161
    toxicity, 8, 161-162, 164, 169, 172-173
  cardiotoxicity, 162, 169
  combination therapy, 158-176
  and infusion, continuous, 158-176
  and radiation therapy, concomitant, 158-176
  radioimmunoassay, 162, 163
  as radiosensitizer, 167-168
Drug, anti-cancer, 3-5
  and cell cycle, 4
  combination, 7
  drug resistance, 4
  pharmacokinetics, 4
  and tumor tissue, 4

Epipodophyllotoxin, continuous infusion, 43-50
Escherichia coli, 177
Esophagus
  cancer, 118-120, 207-209, 217-220
  carcinoma, 139-141
  and cisplatinum, 141
  and FUMIR, 139, see FUMIR
  and metastasis, distant, 141
  and radiation, 207-209
  survival is low, 139
  therapy, listed, 139
Esthesioneuroblastoma, 92, 93
Ethacrynic acid, 55
Etoposide, 9, 44, 211-214
  infusion, continuous, 44-45

Fibrosis, pulmonary, 5, 6, 15
  and bleomycin, 5
FUDR, see Floxuridine
Floxuridine (FUDR) by infusion, 73-78, 195-197
  intrahepatic, 75
  by pump, inplantable, 195-199, see Infusaid
  side effects, 197
  survival time, 197
  systemic infusion, 75
  toxicity, 76
5-Fluorodeoxyuridine monophosphate, 27
5-Fluorouracil, 8, 27-42, 54, 59-62, 107, 113-151, 201-209, 217-220
  action, site of, 30

5-fluorouracil (continued)
  in anal cancer, 120
  in anal canal carcinoma, 135-139
  in breast cancer, 8
  in cervical cancer, 120, 142-143
  clinical status, 116-122
  in colorectal cancer in the liver, 120
  in colorectal carcinoma, advanced, 27-42
  combination therapy, 27-42, 201-205
  and cyclophosphamide, combined, 8
  and DNA, 8, 27, 129
  in esophageal carcinoma, 118-120, 139-141, 207-209, 217-220
  in gastrointestinal cancer, 217-220
  half life very short, 114
  in head and neck carcinoma, 116-118, 141-142
  history, 128-129
  infusion, continuous, 27-42, 107, 129, 201-205
  and leucovorin combination, 27-42
  metabolism, 28-29
  and mitomycin C, combined, 133-147, 201-205
  and mouse, 31-32
  pharmacokinetics, 35-38
  pharmacology, 122
  and radiation, concomitant, 113-125, 133-147, 201-205
  as radiation sensitizer, 113-125, 151
  and rate, 30
  in rectal carcinoma, 120-121
  and RNA, 8, 27, 129
  side effects, 34
  site of action, 30
  in squamous cell cancer, 133-147
  therapeutic index, 28
  and thymidine combination, 27-42
  and thymidylate synthetase, 129
  toxicity, 8, 29, 114
    dermatitis, 115
    stomatitis, 115
5-Fluorouridine triphosphate, 27
Folate cofactor, 27, 35
5-FU, see 5-Fluorouracil
FUDR, see Floxuridine
FUMIR, 133-147
  in anal canal carcinoma, 135-139
  clinical schedule, 134-135
  combined therapy, 133
  toxicity, 134

Gallium, 43
Gastrointestinal tumor, solid, 107-111
Glutamic oxalacetic transaminase, 73

HAI, *see* Artery, hepatic, infusion therapy, *see* Hepatic
Heart failure, congestive, 19-21
  reversible, 24
Hepatic
  artery infusion (HAI), 195-199
  malignancy of childhood, 87-90
  metastasis, 127-132, 195-199
    and rectal cancer, 195
  radiation, 195-199
  resection, 195-199
  toxicity, 73
  see Infusaid, Liver
Hepatoblastoma, 87, 91, 92
  and adriamycin, 87-90
  surgery, 87
Hepatocellular carcinoma, 87
Hepatoma, 165-166
  patient, 56-57
Hickman catheter, *see* Catheter, central venous

Index, therapeutic, 3, 5-6
Infusaid pump, totally implantable, 195-199, 223-228
  cross-section, 227
  and FUDR, 196
  infusion, 195-199, 223-228
  patient teaching, 225-226
  photograph, 226
  procedure for implantation, 223-224
  side effects, 197
  survival time, 197
Infusion, continuous, 3-50, 73, 79-105, 128, 129, 149-153, 177-188, 195-199, 201-205, 211-216

Lactic dehydrogenase, 74
Lemon-Foley chemotherapy, 107-111
  and solid gastrointestinal tumor, 107-111
Leucovorin, 82
  and 5-fluorouracil combined therapy, 27-42
Leukemia, acute, 6
  non-lymphocytic, 95-100
    cytosine arabinoside, drug choice, 95-100
    hospitalization, lengthy, 100
    infusion, continuous, 95-100
    remission possible, 99
Lewis lung carcinoma of mouse, 82
Liver, *see* Hepatic
  cancer, 87-90
  chemotherapy-related complication, 73
  disease, metastatic, 127-132, 195
    and colorectal cancer, 120
    complications, 128

Liver
  disease, metastatic (continued)
    diagnosis, 127
    infusion therapy, 73, 128
    radiation therapy, 128-130
    scan with radionuclide, 74
    treatment, 127
Lymphoma, large cell -, 79-86
  combination therapy, 79
  and COPBLAM, 79
  cure rate, 81

Mandrake root toxin, 44
Matulane, 82
MDS, *see* Myelodysplastic syndrome
Medulloblastoma, 92
  metastatic, 94
  and verapamil, 94
Melanoma patient, 56, 58
Metastasis, hepatic, *see* Hepatic metastasis
Methotrexate, 82
N-Methylene tetrahydrofolic acid, 27
Microsphere and liver cancer, 67-69
  see Starch microsphere
Mitomycin C, 6, 59-61, 68, 113, 118-122, 131-147, 150-151, 201-205, 207-209, 217-220
  and anal canal cancer, 133-147, 151
  and bladder cancer, 151
  and cervix cancer, 142-143
  and esophagus cancer, 139-141, 151
  and head-and-neck cancer, 141-142
  and gastrointestinal cancer, 217-220
  and radiation, concomitant, 133-147, 201-205
  and squamous cell cancer, 133-147, 151
Mucositis, 89, 141, 179, 204
Myelodysplastic syndrome, 95-100

Neuroblastoma, 92
Neuroectodermal tumor, primitive, 91-93
  remission, complete, 93
Neutropenia, 89

Oncovin, 82-83

Paranasal sinus cancer, 190
  case report, 190
Perfusion, regional 3
Podophyllotoxin, 44
  from mandrake root, 44
Porfiromycin, 113
  and radiation, concomitant, 113
Prednisone, oral, 82
Prostate cancer, 155
Pump for chemotherapy
  external, 223-228, *see* Cormed

Pump for chemotherapy (continued)
  implantable, internal, 223-228,
    see Infusaid

Radiation therapy, 113-115, 127-153, 158-193, 201-220, *see* Combined therapy
Radioimmunoassay, 162, 163
Radionuclide liver scan, 74
Radiosensitizer, 113-115, 149, 151, 167, 168
Rectum cancer and hepatic metastasis, 195
Resection, transurethral, 155
RNA, 8, 27, 129
Sarcoma
  and actinomycin D, 183
  and adriamycin, 183-188
  case report, 185-186
  excision, total, 183
  soft tissue, pleomorphic, 183-188
SGOT, *see* Glutamic oxalacetic transaminase
Soft tissue sarcoma, pleomorphic, 183-188
Squamous cell cancer, 133-147
Spherex, *see* Starch microsphere
Spirogermanium, 43
Starch microsphere, biodegradable, 51-66
  and drug delivery in liver tumor, 52
  and hyperthermia, regionalized, 51-66
  and immunotherapy, regionalized, 51-66
  and radiation protection, 51-66
  and serum amylase, 51
  toxicity, 55
Stomatitis, 22, 23

TCC, *see* Carcinoma, transitional cell
Tc99m-Macroaggregated albumin (TcMAA), 68
Teniposide, 44
Thrombocytopenia, 55
Thymidine and 5-fluorouracil combination, 27-42
Thymidylate synthetase, 27
  inhibition by 5-fluorouracil, 129
Tomography, computerized, 74
Tumor
  hepato-biliary, 53
  neuroectodermal, 91, 92
  solid of gastrointestinal tract, 107-111
    case presentation, 109-110
    and 5-fluorouracil therapy, 107-111
    Lemon-Foley method, 107-111
    patient selection, 107
  Wilm's, 91, 92

Uridine triphosphate, 27

Vinblastine, 6, 9
Vinca alkaloid, *see* Vinblastine, Vincristine
Vincristine, 6, 9, 81
  and adriamycin infusion, continuous, 91-94
  and infusion, continuous, 81-83
VM*26*, *see* Teniposide
VP*16-213*, *see* Etoposide

Wilm's tumor, 91-92
  remission, complete, 93

MIX
Papier aus verantwortungsvollen Quellen
Paper from responsible sources
FSC® C105338

If you have any concerns about our products,
you can contact us on
ProductSafety@springernature.com

In case Publisher is established outside the EU,
the EU authorized representative is:
**Springer Nature Customer Service Center GmbH
Europaplatz 3, 69115 Heidelberg, Germany**

Printed by Libri Plureos GmbH
in Hamburg, Germany